Refining Petroleum

SCIENCE in a technical world

A Project of the Education Division of the American Chemical Society

W. H. Freeman and Company
New York

Science in a Technical World Project Team
Co-Principal Investigators: Sylvia Ware, David Lavallee, Kenneth Chapman
Project Director: Ann Benbow
Production Director: Colin Mably
American Chemical Society: Sylvia Ware, Ann Benbow, Janet Boese, Gary Dikeos, Helen Herlocker, Guy Belleman, Michael Tinnesand
Writers: Jack Ballinger, Ann Benbow, Helen Herlocker, Wayne Morgan, Carol Muscara, Marti Tomko, Emmett Wright, Robert Bernoff, Ken Weese
Editors: Marcia Vogel, Ann Benbow, Colin Mably
Project Evaluator: Robert Bernoff
Assessment Consultant: Gail Goldberg
Video Coordinator: Colin Mably
Video Production: Take One Video
CD-ROM Development: CTS, Inc., Sonalysts, Inc.
Safety Reviewer: W. H. Breazeale
Content Advisor: Robert Hofstader

National Science Foundation Visiting Committee
Carlo Parravano, Chair; Barbara Wrobleski, Michael Farmer, Mickey Sarquis, Nancy Dillon, Robert Maleski, Felicia Graves, Toby Horn

W. H. Freeman and Company Book Team
Publisher: Michelle Russel Julet
Sales and Marketing Consultant: Arthur C. Germano
Project Editor: Mary Louise Byrd
Text Designer: Blake Logan
Cover Designer: Patricia McDermond/Blake Logan
Cover Illustration: Roy Weiman
Illustration Coordinator: Lou Capaldo
Illustrations: Proof Positive/Farrowlyne Associates, Inc.
Photo Research: Proof Positive/Farrowlyne Associates, Inc.
Production Coordinator: Julia DeRosa
Production and Composition: Proof Positive/Farrowlyne Associates, Inc.
Manufacturing: RR Donnelley & Sons Company

Library of Congress Cataloging-in-Publication Data

Petroleum refining : a project of the Education Division of the American Chemical Society.
p. cm. — (Science in a technical world)
Includes bibliographical references.
ISBN: 0-7167-4038-9
1. Petroleum—Analysis—Laboratory manuals. I. American Chemical Society. Education Division. II. Series.

TP691 .P454 2001
665.5—dc21 2001023058

Development of the *Science in a Technical World* program was supported, in part, by the National Science Foundation under Grant Nos. DUE 9454564 and DUE 9752102. Opinions expressed are those of the authors and not necessarily those of the Foundation.

Printed in the United States of America
First printing 2001

Science in a Technical World Industry Contacts

ROBERT HOFSTADER (industry consultant); TERRY WEBB, Quality Assurance, Pepsi Bottling Group; JEFF FLYNN, Kleentek; RONALD BARBARO, Northern Virginia Community College; STEVEN CAWTHRON, Leesburg Wastewater Treatment Facility; KENNETH GLANZ, Appleton Papers, Inc.; MARIA SPINU, DuPont Research and Development; PATRICIA LISTER, DuPont Marshall Laboratories; PETER BICK, Eli Lilly and Company; MARK A. HOLCOMB, Dominion Semiconductors; PAMELA MCCARTHY, Civista Medical Laboratory; JENNIFER FERNS, Federal Bureau of Investigation; JERGEN VON BREDOW, Food and Drug Administration; FRED CONFORTH, Phillips 66 Company

Science in a Technical World Pilot Teachers

JEFF ALISAUCRAS, South Carroll High School, Sykesville, MD
JAMES BAUER, McKinley High School, Honolulu, HI
BRITTANY BEHRENS, Coronado High School, Scottsdale, AZ
BARBARA BILLINGS, Milton High School, Alpharetta, GA
BILL CALLAHAN, Watsonville High School, Watsonville, CA
JIM DIXON, Silver Lake Regional High School, Kingston, MA
KATHLEEN DOMBRINK, McCluer North High School, Florissant, MO
RONALD DUNAWAY, Monteray High School, Lubbock, TX
JONATHAN DUNSKI, North Carroll High School, Hampstead, MD
DOUG ENGLAND, Wallace High School, Wallace, ID
ROBERT FOUR-HOGUE, South Carroll High School, Sykesville, MD
JACK GOLDSTON, Ross High School, Fremont, OH
ALAN GONZALES, Lakewood High School, Lakewood, CO
KRISTA GRAY, McDowell Senior High School, Erie, PA
BERNIE HERMANSON, Harlan High School, Harlan, IA
NICK HOFFMAN, Wallace High School, Wallace, ID
JOCELYN HOLLIS, Chester High School, Chester, PA
SAMUEL HOLTZMAN, Pilgrim High School, Warwick, RI
JULIA JANOWICH, Francis Scott Key High School, Union Bridge, MD
CINDY JONES, Jefferson Senior High School, Edgewater, CO
LEAH LINDBLOM, Arvada West High School, Arvada, CO
MARY LINDIMORE, Arvada West High School, Arvada, CO
KAREN LUNIEWSKI, North Carroll High School, Hampstead, MD
DIANNE MCCLEARY, Montwood High School, El Paso, TX
TOM MCCOY, Keystone High School, Knox, PA
SUE MATTESON, McDowell Senior High School, Erie, PA
SHARON MUENCHOW, Arvada West High School, Arvada, CO
ELLIE MUNIZ, Ridgeview High School, Columbia, SC
MIKE MURRAY, Silver Lake Regional High School, Kingston, MA
MARK NALE, Tyrone Area High School, Tyrone, PA
KRISTEN NELSON-MCKENZIE, Thornridge High School, Dolton, IL
BEN NORTON, Bergen County Technical High School, Hackensack, NJ
SCOTT OLSON, Red Rocks Community College, Lakewood, CO
MARIA PACHECO, Buffalo State College, Buffalo, NY
ROBIN PARKINSON, Northridge High School, Ogden, UT
SHELLY PERETZ, Thornridge High School, Dolton, IL
RUSS RAPOSE, Pilgrim High School, Warwick, RI
CANDY RUZIECKI, Chester Senior High School, Chester, SC
BARBARA RYDEN, McDowell Senior High School, Erie, PA
THOMAS SANDHAM, North Kingston High School, North Kingston, RI
WAYNE SILL, Hutchinson High School, Hutchinson, KS
GEOFF SMITH, Watsonville High School, Watsonville, CA
BILL SPEED, Need School, Kailua, HI
TONY TAYLOR, Jefferson County Open School, Lakewood, CO
KATHLEEN THOMPSON, Anthony Jr/Sr High School, Anthony, TX
SCOTT THOMPSON, Arvada High School, Arvada, CO
LILLIE TUCKER-AKIN, Tupelo Christian School, Tupelo, MS
BARBARA WALKER, Ottumwa Community Schools, Ottumwa, IA
LARRY WEATHERWAX, Mclaughlin School, Alaska
KEN WEESE, Montwood High School, El Paso, TX
GARY WHITLING, Keystone High School, Knox, PA
HAROLD WILSON, Medomak Valley High School, Waldoboro, ME
DELORES WRIGHT, Chester High School, Chester, PA
DOUG YUST, Francis Scott Key High School, Union Bridge, MD

Teacher Review Panel

Bernie Hermanson
Nick Hoffman
Mark Nale
Candy Ruziecki
Barbara Walker
Larry Weatherwax
Ken Weese

contents

Multimedia Resources

Refining Petroleum Video: This 13-minute video gives you an overview of the petroleum refining industry at a state-of-the-art facility, as well as the role of technicians in that industry.

Petroleum Refining CD-ROM: The CD-ROM is organized into sections. Petroleum Refining Overview introduces you to the petroleum refining process. Technician Tasks gives more information on the role that technicians play in the refiining process, including preparing samples and running tests for petroleum components, clarity, sulfur content, residual solids, and knocking. Practice Exercises allows you to try laboratory tests and learn procedures. Assessment helps you measure your knowledge and skills relating to the petroleum refining process. Key terms and definitions related to the petroleum refining process are in the CD-ROM's *Glossary* and *Encyclopedia*.

Petroleum Refining on the Internet: A list of Web sites related to petroleum refining, which you will find useful throughout your work in this module, is included at the back of the book.

www.whfreeman.com/STW Additional resources for instructors and students are available online at the *Science in a Technical World* support Web site.

Science Connections Science Connections boxes offer background information and explanations to enhance your understanding of important science concepts covered in this module. You will find them near the places in the book where a key science concept is first introduced.

important notice

Science in a Technical World modules are intended for use by students in the classroom laboratory under the direct supervision of a qualified science teacher. The experiments described in this book involve substances that may be harmful if they are misused or if the procedures described are not followed. Read cautions carefully and follow all directions. Do not use or combine any substances or materials not specifically called for in carrying out experiments. Other substances are mentioned for educational purposes only and should not be used by students unless the instructions specifically so indicate.

The materials, safety information, and procedures contained in this book are believed to be reliable. This information and these procedures should serve only as a starting point for good laboratory practices, and they do not purport to specify minimal legal standards or to represent the policy of the American Chemical Society. No warranty, guarantee, or representation is made by the American Chemical Society as to the accuracy or specificity of the information contained herein, and the American Chemical Society assumes no responsibility in connection therewith. The added safety information is intended to provide basic guidelines for safe practices. It cannot be assumed that all necessary warnings and precautionary measures are contained in the document and that other additional information and measures may not be required.

welcome!

What Are the Goals of *Science in a Technical World*?

Before you begin working with this module, it is important for you to understand what the entire *Science in a Technical World* program is set up to accomplish. Each *Science in a Technical World* module is designed to help you to

- gain an understanding of the **science technology** used in a variety of industries and to become skilled in using science technology to solve problems that actually occur in industry
- become knowledgeable about the **industrial process** in general and compare the employment opportunities that a number of science technology industries offer
- work as a team in a **virtual workplace** setting

What Is Science Technology?

Science technology may be a term that is unfamiliar to you. It refers to the applied science that is used in a wide variety of settings, industry being one.

For example, science technology includes the laboratory equipment and instruments used to monitor industrial processes. These can be the same kind of "wet lab" equipment you may have used in other high school science courses (such as chemistry, biology, or the earth sciences), or they may be very sophisticated, computerized instruments.

This program will give you the chance to learn science technology in an industrial context by letting you and your classmates solve actual problems from different industries.

What Is the Industrial Process?

Industry is the process by which products are made. Some industries with which you may be familiar include the automotive industry (manufactures cars), the food industry (processes food products), the pharmaceutical industry (makes medications), the paper industry, and a number of others.

Science in a Technical World problems are taken from some of these familiar industries, as well as from others that you may not know as much about. The industries selected for the entire program were chosen to allow you to gain science technology skills and knowledge in the biological, chemical, physical, and earth sciences.

What Is the Virtual Workplace?

The **virtual workplace** is a term used in the *Science in a Technical World* program to describe a new way of thinking about your classroom, your teacher, and your classmates. As much as possible, the work that you do throughout the program will mirror what actually happens in industry. This means you will work as teams to solve problems and will be dependent on the work of other teams, at times, to achieve results. You will do research to find information from a variety of sources (including a CD-ROM, a video, and the Internet) to help you solve problems. You will work according to a timeline that all of you establish. You will also keep very careful records of your work, which you will need to have witnessed and signed off on a regular basis. Your teacher's role will shift from an instructor to more of a manager or supervisor of your work.

What Happens After *Science in a Technical World*?

You may decide, after completing the *Science in a Technical World* program, that you would like to learn more about the particular industries that are the focus of the different modules. There are a number of programs that prepare people to work in science technology professions. You might want to investigate these, or contact industries directly to find out what training opportunities they offer.

Whatever you decide, *Science in a Technical World* will give you the knowledge and skills to make an informed choice about careers you may never have considered before. It will also give you a wide breadth of skills that you can use in other science-related fields.

Now, welcome to *this* module, which focuses on the petroleum refining industry.

watch for these symbols

Look for these special symbols as you are reading this book. They have been included to alert you to important learning opportunities, activities, and safety precautions.

Key Question Consider Key Questions carefully. These questions are designed to help you to build your understanding of the concepts, processes, and skills that are central to the focus industry of this *Science in a Technical World* module.

Tips and Hints Tips and Hints offer ideas that will help you along the way in your investigations. They may assist you in answering questions, setting up laboratory equipment, solving problems, or pulling ideas together.

note **Notebook Entry** This symbol reminds you to stop and make an entry in your laboratory notebook. This could be a record of a laboratory procedure, data, an answer to a question, a conclusion, or other important information.

Boldface Terms Definitions are available for terms set in boldface type in the text. They can be found in the Glossary at the end of the module.

MULTIMEDIA SYMBOLS Each *Science in a Technical World* module includes an introductory video, a multipurpose CD-ROM, and suggestions about industry information that can be found on the Internet. The following symbols alert you to opportunities to supplement your laboratory and classroom activities with one of these multimedia resources.

CD-ROM **CD-ROM** This symbol tells you to go to the module's CD-ROM and access one or more of its sections. The CD-ROM might include an Introduction, Technician Tasks, Practice, Assessment, and Reference sections.

VCR **Video** When you see this symbol, you can watch the video to learn more about the production process of the module's focus industry and the role played by technicians in that process.

HTTP: **Internet** This symbol indicates that information that will help you in your investigations is available by accessing the Internet.

SAFETY SYMBOLS Safety precautions are extremely important in laboratory work. Be alert to the safety symbols associated with each investigation in this module, and always follow the safety rules they represent.

 **Safety Goggles** This symbol reminds you to wear safety goggles during the laboratory investigation to protect your eyes from chemical splashes or sudden impacts.

 **Laboratory Apron** This symbol reminds you to wear a laboratory apron during the investigation to protect your clothing and skin.

 **Hot Object or Flame** When you see this symbol, be alert to occasions when you might encounter a hot object (such as a hot plate) or an open flame during the investigation. Secure loose sleeves and long hair before starting your laboratory work.

 **Poisonous Substance** This symbol warns that you will be working with substances that should not be tasted. No substance should ever, under any circumstances, be tasted in a classroom laboratory setting.

 **Special Disposal** This symbol indicates that a substance you will be working with must be disposed of in a special way. You will receive directions on how to dispose of the substance correctly. It is important for safety, health, and environmental reasons that you follow these directions exactly.

 **Dangerous Vapors** This symbol warns you that dangerous vapors might be released in the course of the laboratory investigation. Whenever you see this symbol, your investigation should be conducted under an operating fume hood.

 **Hygiene** This symbol reminds you to wash your hands carefully after a procedure or at the end of the laboratory, as directed.

using your multimedia

Each *Science in a Technical World* module comes with its own video program and CD-ROM.

The video program takes you on a tour of the module's focus industry. You will get to see what products are made, what raw materials are used, and what machinery is important in the industrial process. The video also shows what technicians do in the industry, so you can compare their work with what you do in the laboratory. You may see the complete video at the beginning of the module, or your teacher may show certain parts at appropriate times.

The CD-ROM is a resource for you to use throughout the module. You can access its sections from the main menu by following the voice directions and clicking on the proper screen icons.

The CD-ROM has the following sections:

The **Overview** provides you with information about the focus industry, its products, its processes, and its technicians.

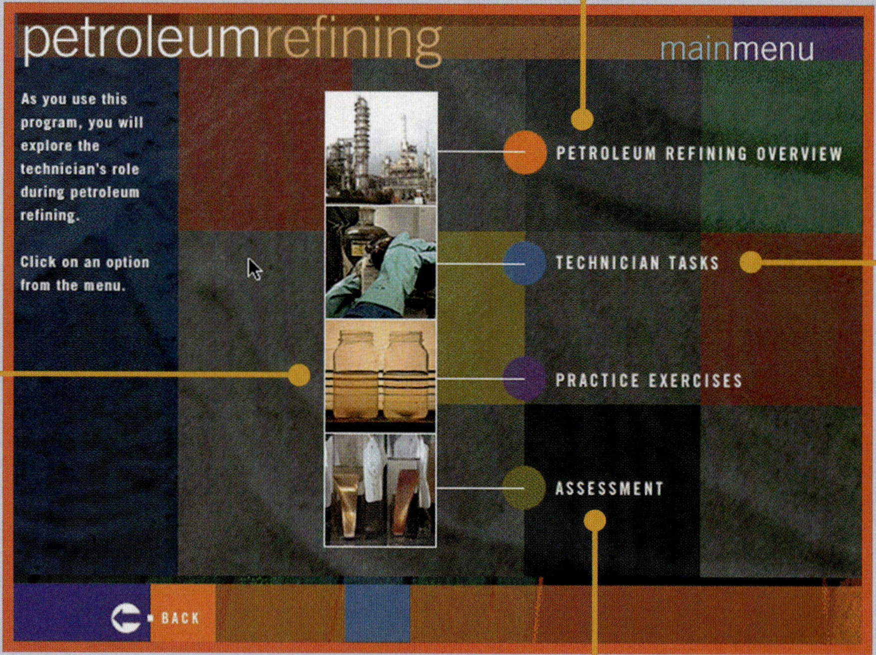

Technician Tasks gives information about what technicians do in the industry. These tasks can include sample preparation, sample checking, sample storage, testing samples, and recording results.

Practice Exercises gives you an opportunity to simulate some of the tests that technicians do in the industry, expand your knowledge of terms and equipment, and manipulate data sets.

Assessment allows you to test your knowledge and proficiency of the tests, terms, procedures, equipment, and calculations that technicians need to know in the industry.

Pull-down menus and icons launch CD-ROM tools:

Clicking **Add to Notebook** will copy the text from the current screen to a special CD-ROM notebook.

Your **Notebook** opens from a pull-down menu. You can edit, save, and print the contents of your notebook. Your teacher may ask you to use your notebook when you are doing the Assessment portion of the CD-ROM.

VCR CD-ROM Watch for **Multimedia Links** throughout this book. They suggest what part of the video or CD-ROM will be most useful to you at each stage of your work with this module.

History tracks where you've been on the CD-ROM. You can navigate to a previous screen by clicking on its name in a displayed list of screens.

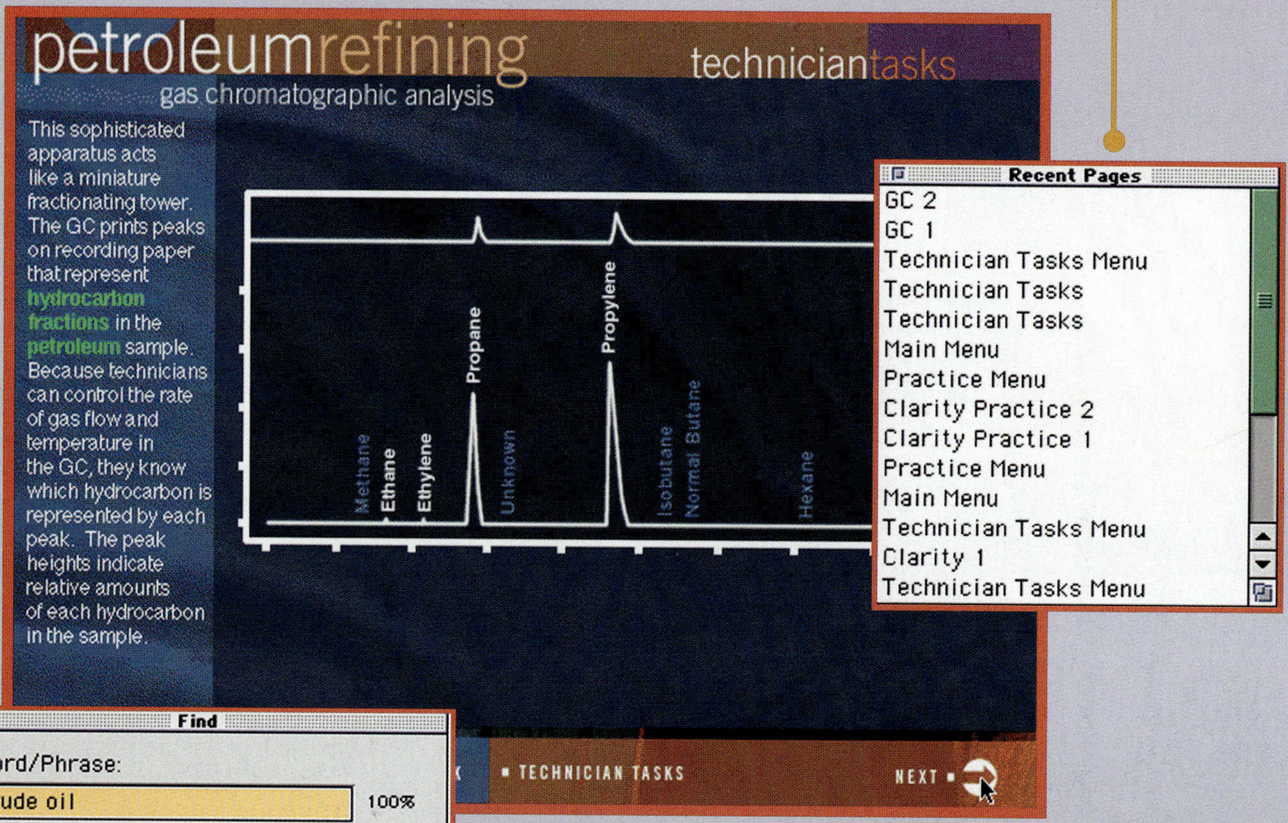

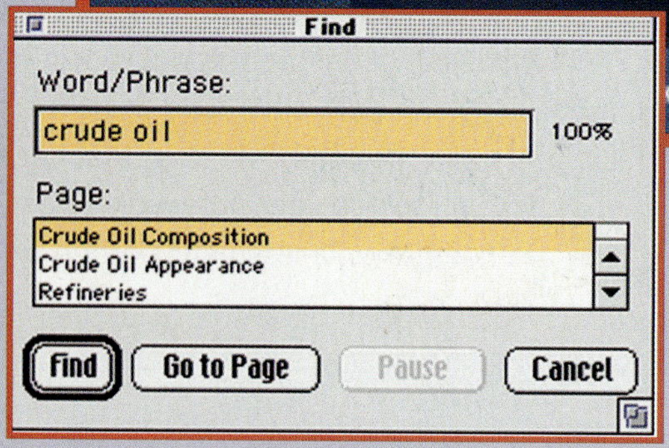

Go allows you to locate a screen by its contents. Clicking on a name in the displayed list of screens takes you to that screen.

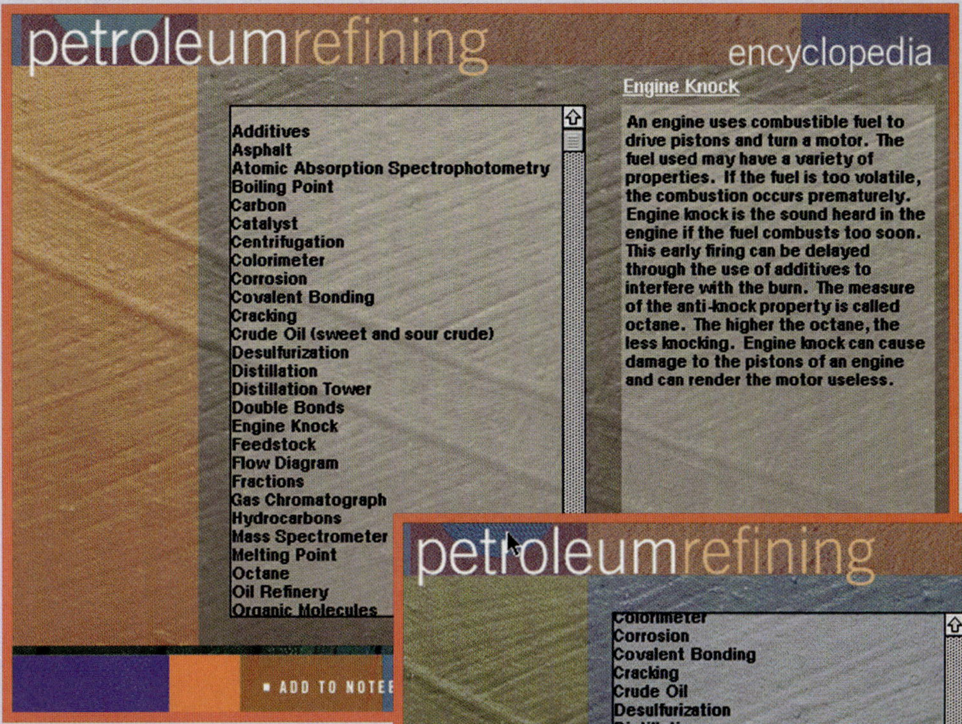

Encyclopedia entries offer more extensive information about selected topics.

The **Glossary** provides definitions of important terms in the module.

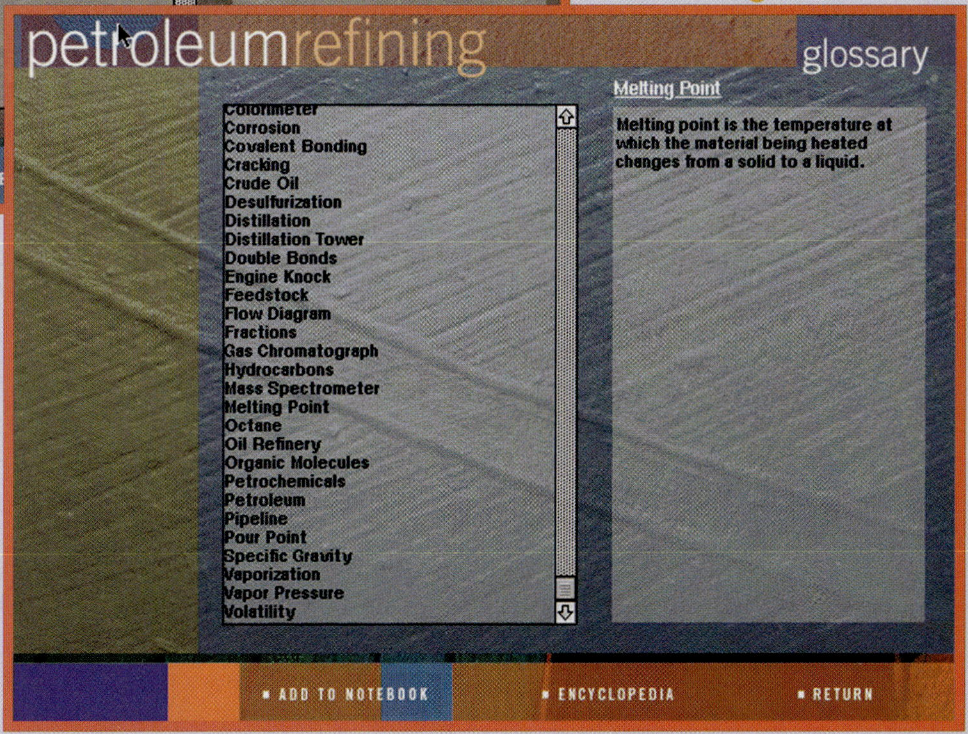

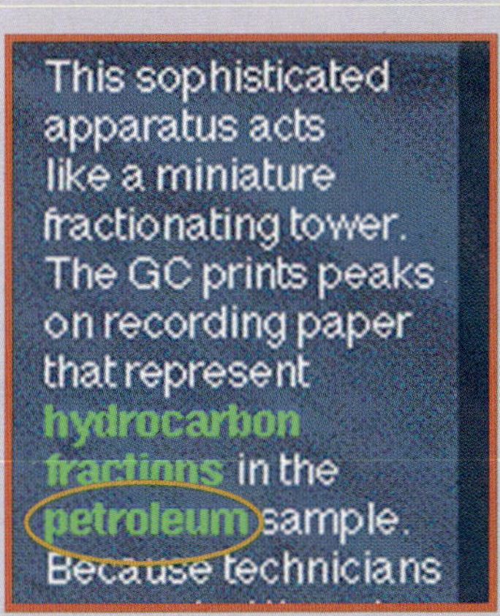

Clicking on a **"hot word"** in the screen text will take you directly to its Glossary definition.

Watch for **Multimedia Links** throughout this book. They suggest what part of the video or CD-ROM will be most useful to you at each stage of your work with this module.

safety in the laboratory

In industry, safety is extremely important. Industries rely on health and safety guidelines set by OSHA, the Occupational Safety and Health Administration, U.S. Department of Labor. These guidelines are tailored to particular industrial processes and practices. In some industries, for example, visitors to a facility must undergo a safety orientation before they are allowed into the plant.

You will have the opportunity and responsibility throughout your *Science in a Technical World* work to observe standard health and safety practices in the laboratory. These practices and guidelines are for your own good; they will ensure that your investigations are rewarding and safe experiences for you and your classmates.

Each laboratory in this book has a Safety section. It is very important that you read and understand the safety precautions before you begin any lab. If you don't understand the safety rules, ask your teacher to explain them. The Rules of Laboratory Conduct listed below apply to *all* laboratory activities included in the *Science in a Technical World* program. Read them now and review them periodically until they become second nature.

Rules of Laboratory Conduct

- Perform laboratory work only when your teacher is present. Unauthorized or unsupervised laboratory work is not allowed.
- Your concern for safety should begin *before* you start a laboratory activity. Always read and think about each assignment before starting it. Ask your teacher to explain any safety rules you don't understand.
- Know the location and correct use of all safety equipment in your laboratory. These should include the safety shower, eye wash, first-aid kit, fire extinguisher, and blankets. Know the location of all exits and proper evacuation routes.
- Wear impact/splashproof safety goggles for all laboratory work.
- Make every effort to protect bare skin. Shorts or short skirts must not be worn; long pants and shirts with sleeves provide better protection. Wear closed shoes (rather than sandals or open-toed shoes).
- Restrain loose articles of clothing, jewelry, and hair.
- Clear your laboratory bench of all unnecessary materials, such as books or articles of clothing, before starting your work.

- Be extremely careful when using sharp tools; never handle their cutting edges, and be sure to store them in an appropriate covering or case when not in use. Also, handle laboratory glassware with care to avoid breakage and the danger of getting cut.
- Check chemical labels twice to make sure you have the correct substance and the correct concentration of a solution (some chemical formulas and names can differ by just a single letter or number). Read and follow the appropriate safe-usage guidelines for each chemical.
- You may, at times, be asked to transfer laboratory chemicals from a common bottle or jar to your own test tube or beaker. Do not return excess material to its original container unless authorized by your teacher.
- Avoid unnecessary movement and talk in the laboratory.
- Never taste laboratory materials. Do not bring food, drinks, or chewing gum into the laboratory. Do not put fingers, pens, or pencils into your mouth while in the laboratory.
- Wash your hands before leaving the laboratory!
- If you are instructed to smell something, do so by fanning some of the vapor toward your nose. *Do not* place your nose near the opening of the container. Your teacher will show you the correct technique.
- Never look directly down into a test tube or beaker; view the contents from the side. Never point the open end of a container toward yourself or a neighbor. Never heat a container directly in a Bunsen burner flame.
- Any laboratory accident, however small, should be reported immediately to your teacher.
- If a chemical spills on your skin or clothing, quickly rinse the area with plenty of water. If the eyes are affected, washing with water must commence immediately and continue for a minimum of 15 minutes or until professional help arrives.
- When discarding or disposing of used materials, carefully follow the instructions provided.
- Return equipment, chemicals, aprons, and protective goggles to the designated locations.
- If in doubt about a procedure or precaution, *ask!*
- Disposable protective gloves must not be reused. The used gloves must be properly discarded.

Module Focus

In this module, you will work as a member of a team to solve a problem encountered in the petroleum refining industry.

Problem to Be Solved

How can laboratory technicians in a petroleum oil refinery organize testing procedures and protocols into flow charts to standardize crude oil sample testing?

introduction

The Problem

You have been hired as a technician at the Gulf Coast Oil Refinery. Although you have worked before as a chemical laboratory technician, this is the first time you have worked with petroleum. As an oil refinery laboratory technician, you will not only need to know the appropriate chemical laboratory procedures for this job, but you will also need to learn the correct and safe procedures for analyzing a petroleum product. Your test results can have a great impact on the petroleum refining process. Inaccurate results could cause useless or even harmful adjustments to the refining process. To prepare for working in the industry, you will participate in an orientation session. The first part of this orientation will focus on the science and technology of the petroleum refining process. This will help you understand the impact your laboratory work can have on the full-scale refining process.

Safety is extremely important in the petroleum refining industry. In the second part of the orientation, you will be introduced to the correct safety, health, and environmental procedures (S/H/E) that apply to the laboratory tests you will be performing. You need to know how to prevent accidents and what to do in case of emergencies. As part of your training, you will develop a S/H/E checklist that you can use throughout your laboratory training sessions.

After completing the orientation session, you will start learning the laboratory work that is part of the essential knowledge and skills for a technician in the petroleum refining industry. You will apply what you learn in the orientation to your laboratory work.

Multimedia Link

CD-ROM

Information on boldfaced terms can be found in the Glossary and Encyclopedia of the Petroleum Refining CD-ROM, as well as in the glossary at the back of this print module.

At the end of the training, you will work with your colleagues to develop a flow diagram for each of the testing procedures carried out in your laboratory. This graphic display of the procedures and protocols will standardize the tests for reliable and consistent reporting of data. By including appropriate safety checklists, you will reinforce the S/H/E procedures mandated for each test.

You will

- examine some sample industry flow diagrams.
- adopt and adapt symbols and icons for your laboratory.
- identify correct procedures for a particular laboratory test.
- identify correct safety, health, and environmental procedures for that test.
- display testing procedures on a flow diagram illustrating the sequence and the key decisions to be made for each test.

Finally, you and your colleagues will organize all of the separate flow diagrams into a master diagram to standardize the sequence of tests to be completed for each incoming sample of crude oil.

It is important to keep careful notes throughout your training. Early in the orientation session you will have the opportunity to conduct research into petroleum refining. Throughout your activities, use a variety of resources to build your knowledge on petroleum and the petroleum refining process.

Your Role in Solving the Problem

You will use a number of models to simulate the processes that technicians use in the laboratory to analyze petroleum samples. Although, for safety reasons, you will not be using the same types of samples used in a petroleum-refining laboratory, you will learn the techniques required to run those tests. A few tests require equipment, such as a gas chromatograph or knock tester, that you probably will not find in your classroom laboratory. In those cases, refer to the Technician Tasks section of the CD-ROM to learn the basics of those tests.

As a laboratory technician trainee in the petroleum refining industry, it is your job to learn selected techniques used in analyzing samples; to understand and use appropriate S/H/E guidelines; and to communicate what you discover from each of your tests. For the purposes of this module, it is also your job to use the knowledge and skills you learn from the module to construct a set of flow diagrams for analyzing crude oil samples.

Fact Sheet on Petroleum Refining

Petroleum products are extremely important to life in our industrialized society. We rely on petroleum products for transportation fuels, heating, manufacturing, and generating electricity. Petroleum is also the source material for many synthetic compounds such as plastics.

Think about what your world would be like without petroleum products. How would you travel any distance? What fuel would you use to heat or cool your home? How would you cook your food? What would take the place of plastics in your life? Without petroleum, we would have to make many adjustments in the way we live.

What Is Petroleum and Where Does It Come From?

Petroleum is pumped from the ground as crude oil, a liquid varying in color from clear to greenish-gray. Some varieties of crude oil are as thin as water; others are as thick as tar. After refining, petroleum is a mixture of compounds composed of the elements carbon and hydrogen. (This is why petroleum compounds are also known as *hydrocarbons.*)

Petroleum is an example of a *fossil fuel.* Over millions of years, the organic remains of marine animals and plants accumulated in layers called *sedimentary layers.* As the number of layers increased, their growing weight generated tremendous pressure and heat—ultimately enough to transform the carbon in the organic sediments into our present-day petroleum deposits. Some places in the world have particularly abundant oil deposits. Regions rich with oil reserves include the Middle East, the coast of the Gulf of Mexico, the northern reaches of Alaska, and the North Sea. Millions of years ago, lush growing conditions favored the accumulation of organic materials that resulted in the formation of petroleum deposits.

Petroleum prospectors use a variety of methods to locate rich deposits. They collect rock samples from an area and search for certain marine fossils, conduct density scans of rock layers, use seismology (an analysis of sound waves passing through rock) to locate possible sites, and examine the actual rock layers. When

they identify an area with the right indicators, prospectors may recommend that an oil company drill a test well. The risks in exploration are high. Most likely, even a site with several positive indicators will not hold enough petroleum—or drilling conditions will be too difficult—to make it financially worthwhile for the company to drill a full-size well.

And even if a petroleum deposit is found, petroleum companies will not drill a well unless the properties of the crude oil are desirable enough to justify the expense. Crude oil from the Middle Eastern countries is relatively easy to reach and its quality is extremely high. This means that Middle Eastern crude oil has far fewer impurities (nonhydrocarbons) than crude oil from other places in the world. High quality crude oil requires fewer refining steps, making it less costly to produce the final petroleum products.

What Is the Chemistry of Petroleum?

Many compounds make up refined petroleum. The base element of all of these compounds is carbon. A carbon atom is surrounded by six electrons. Electrons surround the nucleus of an atom in various energy levels. For carbon, four of its electrons are in the outermost layer of the atom—a layer that may be occupied by a maximum of eight electrons. This means that the carbon atom has four potential sites for sharing pairs of electrons with other atoms. The simplest *hydro*carbon, therefore, is **methane** (CH_4), a carbon atom bonded to four hydrogen atoms. In the structural formula shown here for methane, each line stands for a **single bond,** a pair of electrons shared between the carbon and hydrogen.

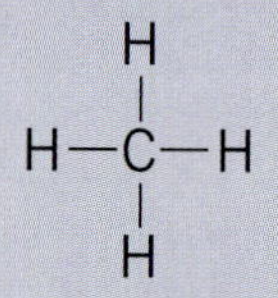

Methane

Two carbons sharing an electron pair with each other can combine with six hydrogen atoms. This is **ethane** (C_2H_6). Here again, each line represents a pair of electrons. The central pair is shared between the two carbon atoms.

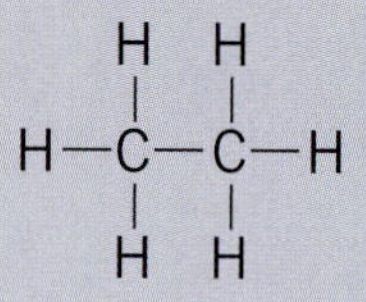

Ethane

Table 1 shows the other, singly bonded hydrocarbons, or **alkanes.** In all alkanes, the carbons are bonded together to form either simple or branched chains that are surrounded by hydrogens.

Other hydrocarbons contain carbons with double bonds (alkenes) or triple bonds (alkynes). These hydrocarbons with double and triple bonds are called **unsaturated** hydrocarbons. The term refers to the fact that the same molecules with single bonds between the carbon atoms could be **saturated** with other electron-sharing atoms, in the case of hydrocarbons, hydrogen atoms.

Name of Compound	Number of Carbons	Chemical Formula
Propane	3	C_3H_8
Butane	4	C_4H_{10}
Pentane	5	C_5H_{12}
Hexane	6	C_6H_{14}
Heptane	7	C_7H_{16}
Octane	8	C_8H_{18}
Nonane	9	C_9H_{20}
Decane	10	$C_{10}H_{22}$
Undecane	11	$C_{11}H_{24}$
Dodecane	12	$C_{12}H_{26}$

Table 1
Singly-Bonded Alkanes

One way in which crude oils are classified is according to the most prevalent type of hydrocarbons they contain. Major classifications include

Alkanes, as shown on page 7.

Alkenes with at least one double bond:

$CH_2{=}CH_2$

Ethene (ethylene)

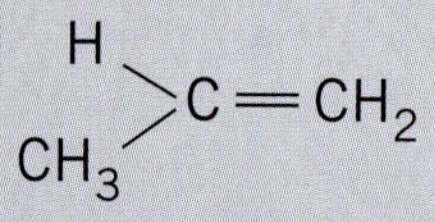

Propene

Alkynes with at least one triple bond:

$HC{\equiv}CH$

Ethyne (actylene)

$CH_3{-}C{\equiv}CH$

Propyne

Aromatics with at least one **benzene** ring, an unsaturated ring that consists of six carbon atoms:

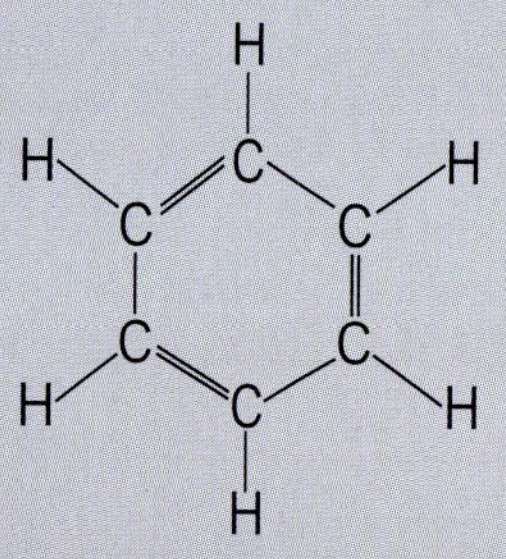

Benzene

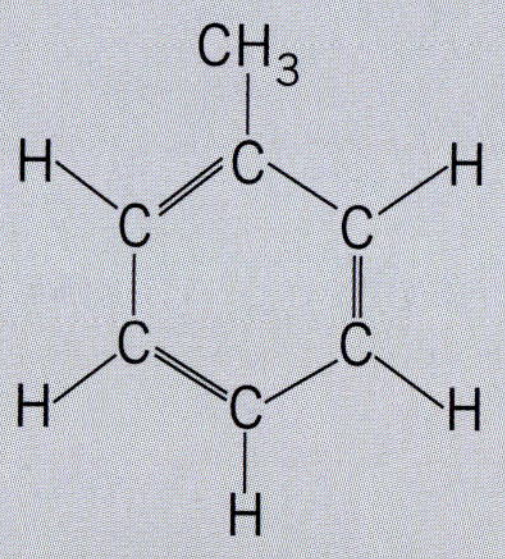

Methylbenzene (Toluene)

The Petroleum Refining Process

Crude oil is a mixture of thousands of individual compounds, most of which are made exclusively from carbon, hydrogen, and oxygen. During refining, the mixture is separated by a process called **distillation** during which the crude mixture is separated into groups of compounds that have similar boiling points. Refinery technicians call each of these groups a *fraction.*

This sequence chain shows the principle steps in the petroleum-refining process and the compounds that are produced at each step:

Crude oil is desalted to remove water-soluble impurities.

↓

Crude oil is sent to the fractioning tower, where the oil is heated.

↓

Different fractions of compounds in the tower boil off at different temperatures (fractional distillation).

↓

Some compounds separated by fractional distillation are further treated by catalytic cracking.

↓

The different fractions are combined with additives and other substances to make products such as gasoline, lubricants, and plastics.

Multimedia Link

CD-ROM

Access the Overview in the Petroleum Refining CD-ROM to learn more about the refining process.

How Is Energy Released from Petroleum Products?

The reaction that releases energy from hydrocarbons is called the **combustion reaction.** In complete combustion, a carbon-based fuel combines with oxygen to release heat, carbon dioxide, and water. Here is the equation for one example of this reaction—the combustion of methane (CH_4), the gas commonly used in gas stoves:

$$CH_4 + 2\ O_2 \rightarrow 2\ H_2O + CO_2 + \text{Heat}$$

Another common combustion reaction is the combustion of octane, a hydrocarbon found in gasoline:

$$2\ C_8H_{18} + 25\ O_2 \rightarrow 18\ H_2O + 16\ CO_2 + \text{Heat}$$

Who Works in the Refining Process and What Do They Do?

Technicians and engineers work together to refine petroleum. Technicians conduct tests on the crude oil to determine its quality. They also monitor the refining process to ensure that the materials produced at each stage meet specifications. Petroleum engineers use the data collected by the technicians to adjust the refining process. Some of the tests that laboratory technicians perform on crude oil (and on petroleum products) include the following:

- specific gravity *(comparison of density of product to the density of water)*
- pour point *(minimum temperature at which a liquid pours)*
- viscosity *(a liquid's resistance to flow)*
- total solids *(the solids content of a mixture)*
- sulfur content *(comparative amount of sulfur compounds in a petroleum-based mixture)*
- clarity
- knock test *(octane rating of a gasoline)*

Developing Your Skills Base

The Petroleum Refining module is designed to help you develop laboratory and problem-solving skills that will be useful to you throughout the *Science in a Technical World* program and throughout your life. You will have many opportunities to practice and refine these skills both in the laboratory and in the virtual workplace found on the CD-ROM. It will be up to you to monitor how well you are mastering a particular skill and to get help when you think you need it.

- *Throughout the module, pay special attention to how you observe, measure, and test for the properties of different materials. The ability to make* ***quantitative and qualitative observations*** *is a vital skill for laboratory technicians.*

- *Problem solving is also an essential technician skill. Learning problem-solving strategies and science processes will prepare you for solving not only petroleum-refining problems, but also other science and technology problems in the future.*

important concepts, processes, and skills

Science Concepts

When you finish this module, you should be able to *define* and, where appropriate, *give examples of*

- hydrocarbons
- the components of crude oil
- the differences between the various types of crude oil
- the molecular structures of selected components of crude oil
- organic and inorganic components of crude oil
- the petroleum refining process
- important physical properties of crude oil
- combustion of a hydrocarbon

Multimedia Link

CD-ROM

Information on each of these Science Concepts can be found in the Reference section of the Petroleum Refining CD-ROM.

Science Processes

When you finish this module, you should be able to

- *define* problems to be investigated
- *generate* and *test* questions and hypotheses
- *construct* a valid experimental test
- *gather* and *record* data
- *organize* data into charts and graphs
- *analyze* data
- *arrive* at conclusions based on data
- *communicate* results
- *connect* work performed to the goals of the industry

Laboratory Skills

When you finish this module, you should be able to

- *prepare* and *store* samples safely
- *follow* directions for an experimental test
- *ignite* a laboratory burner
- *calculate* percentage by volume of water in a mixture
- *conduct* a pour point test for viscosity and purity
- *measure* mass using a balance
- *conduct* a test for copper corrosion
- *measure* volume using volumetric glassware
- *set up* and *use* a distillation apparatus
- *set up* and *use* a centrifuge
- *conduct* a flame test
- *find* the density of a liquid
- *measure* the specific gravity of a liquid
- *set* a calibration standard
- *follow* appropriate safety procedures

Workplace Skills

When you finish this module, you should be able to

- *keep* an accurate log of all laboratory work
- *read, understand,* and *follow* written directions
- *read* and *interpret* a Material Safety Data Sheet
- *follow* safety/health/environmental guidelines
- *devise* and *use a checklist* for safety/health/environmental guidelines
- *anticipate* the work load and *meet* deadlines
- *access* computers and other resources as necessary to find information
- *work cooperatively* with team members
- *devise* and *test* laboratory flow-charts
- *develop* and *deliver* a clear and accurate presentation of results

Getting Started on the Problem

Before you can begin acquiring the technical skills and knowledge for working in the petroleum-refining laboratory, you need to become familiar with the industry and with the roles of laboratory technicians and engineers who work there. To do this, you will begin by learning what petroleum is, where it comes from, how it is separated into useful components, how the components are used, and how they are tested at various stages in the laboratory. It is important to identify desirable properties for individual petroleum products, and to test samples for these properties. Be especially aware of all links between your laboratory work and the impact of this work on the full-scale oil refining process. Finally, remember that a major goal of your training will be to understand and apply the appropriate safety precautions for each test you do. All of these aspects will be part of the final assessment at the end of the module.

Before attending any training, running any tests, or doing any research, you need to set up a laboratory notebook. A technician must have a place to keep a permanent project record of all test procedures and findings. The first activity will show you how to set up your notebook so it is easy to use and to read.

Step 1 Setting Up and Maintaining a Laboratory Notebook

Materials (for each team member)

- bound notebook
- pen

Legal Issues

In the workplace, the laboratory notebook is an important legal document that may be used to witness, verify, and validate findings. In most laboratories, the technician must sign each page, attesting to its accuracy. Records in laboratory notebooks document all laboratory tests done for a company. Considered to be legal documents, such records may be important evidence in court proceedings related to patent law. To be legally acceptable, laboratory notebooks must be completed and signed by technicians daily. Each page must be

- numbered
- dated
- witnessed by another person

Because others may need to replicate your experiment or procedure, it is very important that your laboratory notebook be neat, accurate, and written in pen. Pages should be dated so that it is easy to follow the sequence of your experiments or tests. In addition, record any unique or new event with a note of the exact time the event took place.

Procedure (for 3 or 4 students)

Each team member should keep his/her own laboratory notebook. It should contain a record of all

- predictions
- tests and procedures
- data
- observations
- calculations
- notes
- conclusions

Technicians follow standard procedures for tests. The standard test that is used to measure a particular property *must* be recorded in the laboratory notebook. You may also be required to include a description of each complete experiment and a summary of all tasks completed by your team.

At the end of each class period, remember to have your notebook witnessed and signed off by another member of your team.

Handling Data

Data must *always* be labeled and described. Be sure to record

- the identity of every sample that is tested
- the time of day the test was run
- environmental conditions (temperature, humidity) at the time the test was run
- your observations about the samples

Including observations about samples as part of your data is particularly important in the oil refinery laboratory. Observations about petroleum product samples will include such characteristics as color, odor, temperature, and clarity.

What you write about one sample distinguishes it from any other.

A table is an effective way to present data, as it is easy to review later on.

Connecting to the Problem

Although a procedure and its results are very familiar to a technician while he or she is running a test, later it can be hard to remem-

Figure 1
Among the records that you may record in your lab notebook are the storage temperature for samples.

ber just what a sample looked like or what results the test produced. This is why a well-kept laboratory notebook is necessary—it serves as a complete history of every raw material, process, and experimental result. No information is unimportant (Figure 1).

Maintaining a laboratory notebook also helps technicians make sure that a product meets company specifications, ensuring that customers consistently receive the best possible product.

Only clear, accurate records of data and procedures are acceptable for a technician's laboratory notebook.

Multimedia Link

CD-ROM

In addition to your laboratory notebook, you will have an electronic notebook on the CD-ROM. You can copy text from the CD into this notebook, edit it, print it, and save it to a floppy disk.

Step 2 Learning About the Petroleum Refining Process

Activity Overview

Before you can conduct laboratory tests to monitor the petroleum refining process, you need to know more about what petroleum refining is. You will be using all the resources at your disposal—

Industry Overview, print reference materials, CD-ROM, video, and Internet—to find out about both petroleum and petroleum refining.

Multimedia Link

CD-ROM Access the CD-ROM to find the section entitled Overview. It contains background information on petroleum and the refining process. You can also look up terms in the glossary and encyclopedia found in the reference section of the CD-ROM. Text from the Reference section can be copied into your electronic notebook.

Materials

- Refining Petroleum CD-ROM
- Refining Petroleum video
- petroleum web sites (see Additional Resources at the back of this book)
- petroleum trade publications
- books and journals on petroleum and refining
- laboratory notebook

Safety

No special safety precautions are needed for this activity.

Procedure (for 3 or 4 students)

1. You will use all the resources in your classroom to find out as much as you can about petroleum and petroleum refining. These questions can help you focus your research:

- What is petroleum?
- What are the components of petroleum?
- Why is petroleum such a valuable resource?
- Where on Earth is petroleum found?
- How is petroleum found?
- How was petroleum formed in the first place?
- How is petroleum extracted?
- What processes are needed to make crude oil useful?
- What are the important properties of key petroleum products?
- What are the hazards in working with petroleum?
- What reactions are involved in releasing energy from petroleum products?
- What are the environmental impacts of using petroleum products?

2. Divide up the work between team members so that each person is responsible for particular topics or areas of interest. As you record your findings, be sure to cite your sources so that others can verify what you have discovered.

3. Start your research by rereading the Industry Overview. Although it contains extensive information about petroleum and petroleum refining, you will find that you need to do additional research to answer all the questions in Step 1.

4. As you carry out your research, organize your information in the most effective way you can. Share your findings with the rest of the class in written reports, posters, multimedia presentations, or models and displays.

Analyzing the Information

Consider what the entire class has discovered about petroleum and the refining process. In your notebook make a list of the things you now understand about petroleum refining; then list any questions that you still have. Consult with others about your questions—they may be able to direct you to a useful source of information. note

Arriving at Conclusions

Working with your technician colleagues, make a list of the industrial roles of a petroleum oil refinery. Include the types of raw materials, processes, and products that the industry includes. Identify ways that oil refining impacts other industries. Try to locate the role of an oil refinery laboratory technician in this big picture of the industry. List these ideas, and post them in the classroom for frequent review and revision as you become more familiar with the oil refining industry.

Throughout the module, continue to build your understanding of petroleum and the petroleum refining process.

Connecting to the Problem

Go to the Overview section of the CD-ROM and assess your current understanding of the petroleum refining process. In your electronic notebook, record what you already know and what you still need to find out about petroleum refining. note

Step 3 Safety Guidelines for Petroleum Refining

Since crude oil and the products produced by refining crude oil are highly flammable as well as toxic, attention to safety, health, and the environment is absolutely crucial in an oil refinery. Technicians must wear the appropriate personal safety items and follow S/H/E mandates as they work. Such mandates include following safety protocols in the collecting, testing, and disposing of samples.

As part of your orientation to petroleum refining, you need to understand the need for S/H/E guidelines in the laboratory. Every technician working in the laboratory is responsible for ensuring that he or she understands S/H/E requirements and follows these for each test performed. Accidents put both employees and products at risk. It is common practice for petroleum companies to have comprehensive S/H/E plans at each of their facilities. These plans include some or all of the following components:

- employee training
- stringent S/H/E regulations in force
- well-maintained equipment
- correct and well-maintained safety equipment
- proper means of disposal
- recording and monitoring S/H/E compliance
- team plans for achieving compliance (and going above and beyond)
- quick response to accidents
- community involvement

As you progress through the safety training, then through the analysis laboratories, be alert to the S/H/E guidelines for each laboratory, and follow them carefully. At the end of this part of the training, you will be asked to evaluate the guidelines and to make suggestions for improvements.

Part A Reading and Interpreting Material Safety Data Sheets

Activity Overview

In this first activity, you will learn to read and interpret a Material Safety Data Sheet.

Background

Manufacturers of chemical materials are required by the Occupational Safety and Health Administration (OSHA) regulations to have a Material Safety Data Sheet (MSDS) available for each of their products. Each MSDS contains information about the physical and chemical properties of the material as well as guidelines for handling, disposal, and other procedures. note

First, you will learn to interpret an MSDS for a common household liquid—mouthwash. Next, you will use this information to devise a safety plan for handling the mouthwash as you separate its components by performing a distillation procedure. By the end of the work, the class will have produced several MSDSs, representing different brands and manufacturers of mouthwash. You will specialize in one type of MSDS, then exchange information with other teams in class.

Materials for the class

- several MSDSs for different brands of mouthwash
- samples of mouthwashes with labels
- computer with Internet capacity

Safety

No special safety requirements are needed for this activity because the mouthwashes remain in closed containers throughout.

Procedure (for 3 or 4 students)

1. Read over the MSDS for your mouthwash. Work with your team to find the information requested in Table 2 below. Note: You may need to add columns for additional ingredients.

Table 2
Mouthwash Characteristics

Name of Mouthwash: **Manufacturer:** **Date of MSDS:**			
	Ingredient 1	Ingredient 2	Ingredient 3
Physical and chemical properties			
Fire and explosion hazard data			
Reactivity data			
Health hazard data			
Precautions for safe handling and use			
Disposal			

2. Make a copy of the table in your laboratory notebook. Fill in the table for your mouthwash as completely as possible.

3. Exchange information with other groups in your class. Make copies of completed tables so that everyone has a copy of everyone else's MSDS data.

Analyzing the Data

1. What do all of the mouthwashes seem to have in common?

2. How easy was it to find information on the MSDSs? Do they all follow the same format?

3. What are the recommendations on the various MSDSs for handling the mouthwashes?

4. What are the guidelines on the MSDSs for disposing of the mouthwashes? Do these make sense when you think of how the mouthwashes are used in homes?

Connecting to the Problem

Technicians working in industries use MSDSs as guides for designing safety plans and procedures for handling and disposing of materials. In the petroleum refining industry, where all of the products are toxic and pose fire hazards, MSDSs are crucial in providing accurate information for effective safety plans.

There are a number of sources for MSDSs on the Internet. Using a search engine, key in the words: Material Safety Data Sheets. You will get a list of databases containing thousands of MSDSs for products. If you have the appropriate software (available on the Web for free download), you can print out some of the sheets. Another way to obtain MSDSs is by contacting manufacturers directly through a product label address or toll-free number.

Part B Using Material Safety Data Sheets as a Basis of S/H/E Guidelines for a Distillation Laboratory Using Mouthwash

Activity Overview

In this activity, you will use information from Material Safety Data Sheets on mouthwashes to check and enhance safety guidelines for a distillation laboratory procedure.

Background

Distillation is a process used to separate the components of a liquid mixture. Each component of a mixture has a characteristic boiling point. This is one of a substance's *physical properties.* As you are probably aware, water has a boiling point of 100 degrees Celsius (°C) at sea-level atmospheric pressure. Other substances in a mixture may have boiling points higher or lower than this. Boiling point is very useful in separating mixtures of liquids; as the mixture is heated, each liquid substance begins changing to a gas when the temperature reaches its unique boiling point. In a distillation apparatus, the mixture to be distilled is first heated in a flask. As each fraction (different substance) evaporates, the gas enters a **condenser,** where chilled coils cause the gas to change back into a liquid, or **condensate.** With care, the condensates of the substances can be collected, one after another.

The distillation process is used in petroleum refineries to separate the liquid components of a crude oil mixture into useful fractions (see Figure 2). These fractions undergo further treatment and processing to yield new products.

In this laboratory, you will use the information from the MSDSs for mouthwash to critique and improve the safety plan for the distillation of mouthwash.

Materials

Assemble all of the materials for Laboratory 1: Standard Distillation. See pp. 26–27.

Safety

See the guidelines for handling glassware in Step 4.

Figure 2
At a refining facility, distillation procedures are carried out in a structure called a fractioning tower.

Procedure (for 3 or 4 students)

1. Read over Laboratory 1 (pp. 25–30). In your notebook, record questions that you have on the safety, health, and environmental factors that must be considered in performing the laboratory. Share your questions with your team.

2. Set up the distillation apparatus, being sure to follow safety guidelines for handling glassware and thermometers. However, *do not* turn on the heat or put any liquid into the flask.

3. With your team, examine the assembled apparatus critically. Note all possible hazards. Once the distillation begins, where could things go wrong? Using the diagram of the distillation apparatus as a guide, mark places that should always be checked before a distillation begins.

4. Check the status of each piece of equipment and of your safety equipment. Ask yourself these questions:

- Is any of the glassware cracked?
- How will the flask be heated? Is the heating apparatus capable of providing a steady, diffused source of heat?
- What is the condition of the tubing? Does any need to be replaced?
- What is the condition of the safety equipment in the laboratory?
- Does the fume hood work?
- How is the laboratory set up to handle the types of accidents that could occur during a distillation?

5. Review your MSDS information tables on the mouthwashes, and answer these questions.

- Why is it important to be sure that the apparatus heats gently?
- What precautions should you take when handling the mouthwashes and the fractions taken from them?
- How should you dispose of unused mouthwash and the fractions?

6. When you have finished examining the assembled apparatus and reviewing the MSDS data, put together an improved S/H/E plan for the distillation laboratory. Make sure that it covers all of the critical points. Have you included the following cautions?

- Proper personal safety protection
- Equipment checks
- Safety equipment availability
- Accident plan
- Disposal
- Handling

- Recording of safety procedures
- Other important points

Analyzing the Data

Share your plan with other teams in your class. Ask questions to clarify plans suggested by other teams.

Arriving at Conclusions

After analyzing everyone's plan, develop an overall S/H/E plan that everyone can use for the distillation procedure. Ask your teacher to approve the plan. Post a copy of the overall plan so that everyone is aware of what to do.

Connecting to the Problem

S/H/E procedures must be a team effort in industry. Part of your responsibility is to provide input to the ongoing S/H/E process and to follow the regulations, for your own safety and for that of your colleagues and the community.

Make a checklist of the key points that go into writing a S/H/E plan for any laboratory procedure. You will be using this checklist to plan for each of the laboratories in your training session. You may need to revise the checklist as you proceed.

Now that you have finished your orientation sessions in petroleum refining content and safety, your next task is to learn the laboratory tests and procedures you will perform as a technician. As stated in the section Your Role in Solving the Problem (p. 3), many of these tests model the basic procedures without actually using the hazardous materials used in the petroleum refining industry.

It will be important to use your S/H/E checklist for each of the laboratories in order to write a S/H/E plan similar to the one for Laboratory 1. You will also need to connect what you do in each of the laboratories with the full-scale petroleum refining process. At the end of the training session, you will be assessed on both your knowledge and your laboratory skills to determine your level of proficiency as a petroleum-refining technician.

Multimedia Link

CD-ROM

Access the Technician Tasks section on the CD-ROM as you work through each of the laboratories in this part of the training. The Technician Tasks will help you understand how the activities in the laboratory affect the full-scale petroleum refining process.

Laboratory 1 Standard Distillation

Laboratory Overview

In this laboratory you will learn how to set up a **distillation** apparatus and will distill a mixture (mouthwash). Distillation is one process used to separate the mixture of compounds that is crude oil.

Background

Crude oil is a mixture of liquids. After technicians remove a variety of water-soluble salts by a method called **desalting**, the remaining mixture contains liquid **hydrocarbons**—molecules containing only two elements, hydrogen and carbon. Separating these molecules is the goal of the refining process.

Consider this problem. Imagine that you have been given the task of separating a carload of paper clips, pencil shavings, and table salt, all mixed together. To complete your task, you can probably devise a number of strategies involving magnets, burning, sifting, washing, and the like. The strategies you imagine are based on your knowledge of the unique properties of each item in the mixture.

Science Connections
Distillation in the Petroleum Refining Process

The crude oil pumped from underground reservoirs and transported to distant points for processing represents an energy-rich concentrate that surpasses all other organic fuels in potential yield. However, unlike natural gas, crude oil is actually a mixture of hundreds of molecular compounds. As for any mixture, the strategies for separating these compounds are based on the different properties of the components.

A boiling mixture behaves differently from a pure sample. A pure sample boils at a constant temperature until all of the liquid has been converted to gas. For example, at sea-level atmospheric pressure, pure water boils at a steady temperature of 100 °C, while a mixture of liquids boils over the range of all of the boiling points of the various liquids in the mixture.

As you learned in the Activity Overview to Part B (see p. 20), distillation is a strategy for physically separating a mixture of liquids. First, the mixture is heated. Since each liquid in the mixture has a unique boiling point, each fraction will vaporize at its own point in the rising temperature sequence. The resulting gaseous fractions are quickly cooled and liquefied for collection. Fractions with low boiling points are said to be more volatile than those that have a higher boiling point. Thus, lighter, more **volatile** molecules such as propane boil off first, followed in order by the heavier, less volatile substances, like diesel fuel oil. At the end of the distillation process, asphalt is all that is left.

Distillation only separates the original mixture of liquids. No chemical changes occur during the process.

Materials

- safety goggles (1 pair per student)
- aprons (1 per student)
- 1 alcohol thermometer, maximum temperature 110 °C
- 1 two-hole stopper to fit Erlenmeyer flask
- 1 250-mL Erlenmeyer flask
- 1 condenser (design and scale will determine the glass tubing needed to connect to flask)
- 2 lengths of flexible tubing for connecting condenser to cold water and to drain
- 1 hot plate
- 25 small containers for collecting condensate (small medicine cups or beakers), each capable of holding about 5 mL of liquid

- 1 10-mL graduated cylinder
- 1 tube stopcock grease, shared with other groups as needed
- 10 boiling chips
- heavy cloth towel
- 100 mL of sample mixture (mouthwash) plus an additional 5 mL set aside in a sample cup marked "Original"
- waterproof marker
- MSDS and labels for mouthwashes
- S/H/E checklist from Part 1

Safety

Consult the S/H/E plan you made in your safety training and the information below:

- Wear safety goggles and a laboratory apron throughout the procedure.
- Use the correct procedure for inserting the thermometer and tubing into the rubber stopper.
- Remember, tasting is never appropriate in the chemistry laboratory.
- Use caution when smelling a liquid. Hold the cup several inches away from your nose and waft the vapors toward you by fanning them with your hand.
- Have your teacher check the distillation apparatus before turning on the hot plate.
- Work under a hood.
- Do not let the sample boil dry.
- Keep samples and condensate cups away from any open flames.
- Use caution when you are handling the distillation apparatus. Remember—hot glassware does not look hot.
- Dispose of used solutions as your teacher directs.

Procedure (for 3 or 4 students)

1. Pour 5 mL of water into one small cup, and mark the level with a permanent marker.

2. Duplicate this mark at the correct height on the remaining sample cups.

3. Number the cups 1–25.

4. Obtain one more cup and mark it "Original." Use it to obtain a sample of the mouthwash, and note the substance's physical properties of color and odor.

Figure 3
Note the details of the setup for the distillation apparatus.

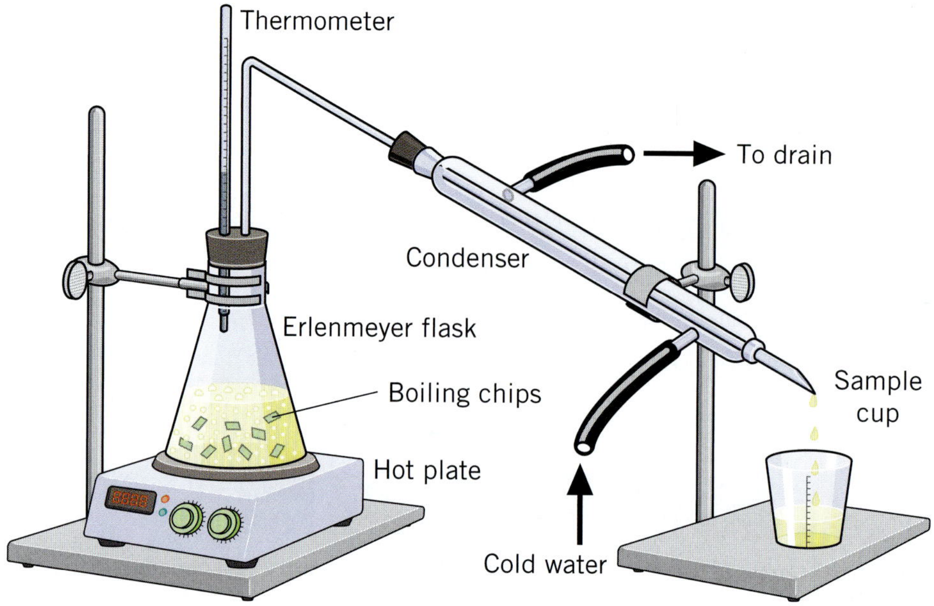

5. Apply a small amount of stopcock grease to the tip of the thermometer and to any glass tubing to be inserted through the rubber stopper. Holding the glass tubing and the stopper with a heavy cloth towel, insert the tube as directed by your teacher.

6. Assemble apparatus as shown in Figure 3 placing 100 mL of mouthwash in the flask. Have your teacher check your set-up before proceeding to step 8.

7. Turn on the water.

8. Turn the hot plate to "MEDIUM." Heat the mixture to 80 °C. Reduce heat to a "MEDIUM" setting.

9. Begin sample collection and data recording at 80 °C.

Temperature Data Collection

Set up a table for recording the temperature. You will begin taking measurements when the first fraction begins to collect. The last item that you record should be the temperature when approximately 5 mL of sample remains in the flask. Do not allow the mixture in the flask to heat to dryness. Note: Table 3 is an example. You may need to add more rows for your fractions.

Sample Collection

1. Place a cap at the end of the condenser before the distillation is started.

2. Record the temperature when the first drop of liquid enters the cup.

3. When 5 mL of liquid has been collected, record the temperature, and quickly replace cup 1 with cup 2. When only 5 mL of

Table 3
Observations of Mouthwash Fractions

Fraction #	Temperature (°C)	Volume (mL)	Odor	Appearance
Original:				
1.				
2.				
3.				
4.				
5.				
6.				
7.				

liquid is remaining in the distilling flask, be certain to turn off the hot plate.

4. Continue recording and replacing cups until only about 5 mL of sample remains in the flask.

5. Observe and record the volume, appearance, and odor of each fraction.

Analyzing the Data

The collected samples represent fractions of the original mixture separated according to the physical property of boiling point.

- Do these fractions differ by any other properties?

Examine the cup marked "Original," the row of numbered samples, and the liquid that remains in the flask.

- How do the fractions differ from the original liquid?

Look at the temperature information for each numbered sample that you recorded in Table 3.

- Which samples are the most similar in terms of temperature of collection?
- Which samples differ the most in temperature of collection?

Examine the properties for each sample that you recorded in Table 3.

- If two samples have similar properties, were they collected at similar temperatures?

Test the relationship between samples with similar properties in several ways. Group the cups, and check the temperature data. Compare two cups with widely spaced temperatures of collection.

- Are the substances similar or different in physical properties?
- What is the total volume of the cups collected at similar temperatures (grouped fractions)?

Arriving at Conclusions

Read the MSDS and labels for the mouthwash you distilled. Product labels list the ingredients in order from most abundant to least. Think about grouped fractions you collected.

- Based on their relative volumes and properties, what do you think your grouped fractions are?
- What is the evidence for your identifications? Where did you find this evidence?
- What tests would help you make more conclusive identifications?

Connecting to the Problem

Distillation is one strategy used in the petroleum refining industry for classifying the crude oil as paraffinic, naphthenic, aromatic, or mixed. These classifications refer to the predominant type of hydrocarbon in the crude oil. See the Industry Overview at the beginning of this module for more information on the hydrocarbon classifications of crude oil.

- How do you think a technician uses distillation in analyzing a sample of crude oil?

At the conclusion of your technician training, you will include this distillation procedure on your laboratory flow diagram. note

Multimedia Link

CD-ROM In the Technician Tasks section of the CD-ROM, you can learn about another tool that technicians use to detect various components of crude oil: a gas chromatograph, which vaporizes a sample and separates it into its components.

- How does the distillation process reflect the refining process? Use the video and the CD-ROM to help you explain.
- How do you think a distillation process could be used to help identify the components of a sample of crude oil?

Laboratory 2 Standard Test for Density

Overview

In this training laboratory, you will learn how to calculate the densities of several sample liquids and rank them from lowest to highest.

Background

The **density** of any substance is defined as the mass per unit of volume. The units you will use in this laboratory will be grams (g) per milliliter (mL). The density of water—1.00000 g/mL at 3.98 °C—is your technical standard for comparison in both of the laboratory procedures. Tables in chemistry handbooks provide information about water density for a range of temperatures. However, for the purposes of your laboratory tests, you can accept that the density of water is 1.0 g/mL at normal room temperature.

Density is a physical property that can be used to describe and identify pure substances. When petroleum products are blended, the blended mixture also has a characteristic density. To analyze a mixture, the technician determines the density of the mixture's most plentiful component. Adding other components to the mixture either raises or lowers this density. Using this information, technicians can test for the relative amounts of different products in the mixture.

Materials

- safety goggles (1 pair per student)
- aprons (1 per student)
- 10 identical disposable 3-mL micropipettes
- 1 set of 5 pure liquids to be tested, 100 mL each, at room temperature
- MSDSs for the 5 test liquids, names masked
- 100 mL distilled water, room temperature
- 1 balance, readable to 0.01 g
- 1 250-mL beaker for supporting micropipettes
- 1 marking pen
- 1 5-mL graduated cylinder
- chemistry handbook
- table for recording data
- S/H/E checklist

Safety

Using your S/H/E checklist, prepare a S/H/E plan for this procedure. Be sure it includes the following guidelines:

- Wear safety goggles and a laboratory apron throughout the procedure.
- Remember, tasting is never appropriate in the chemistry laboratory.
- Be careful when handling glassware.
- Dispose of used solutions as your teacher directs.

Table 4
Density Table

Liquid Sample or Fraction	Weight of Empty Pipette (to nearest 0.01 g) (a)	Volume of Pipette (to nearest 0.1 mL) (b)	Weight of Sample + Pipette (to nearest 0.01 g) (c)	Weight of Sample (to nearest 0.01 g) [c – a = (d)]	Density Sample (g/mL) (d/b =)
Water standard					
Samples					
(Add rows to table as needed)					

Procedure (for 3 or 4 students)

1. Read the entire laboratory activity and the MSDSs for the test liquids. (These sheets will be coded to the numbers of the test liquids, but will not have the names of the liquids on them.) Consult your S/H/E plan for this procedure before you begin. Make sure that you understand what you need to do and what precautions you need to take.

2. Construct a table for recording your data similar to the one above.

3. Number the micropipettes using the marking pen.

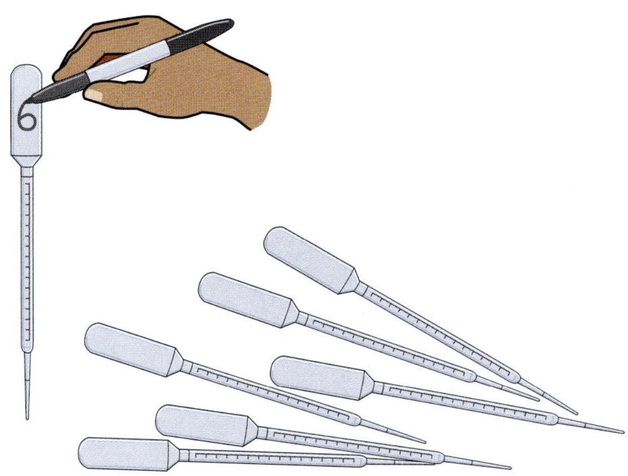

4. Weigh each pipette and record its weight to the nearest 0.01 g in your laboratory notebook.

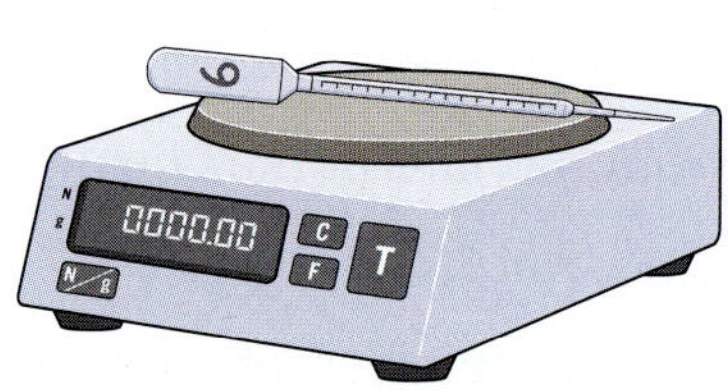

5. Select the six pipettes that have the most consistent masses. Use these for holding sample liquids.

6. Practice filling one of the pipettes with water until you devise a method for filling the pipette as full as possible. Then empty the full contents of the pipette into a 5 mL graduated cylinder, and record the capacity of the pipette to the nearest 0.1 mL. You will use this volume for calculating densities in step 11.

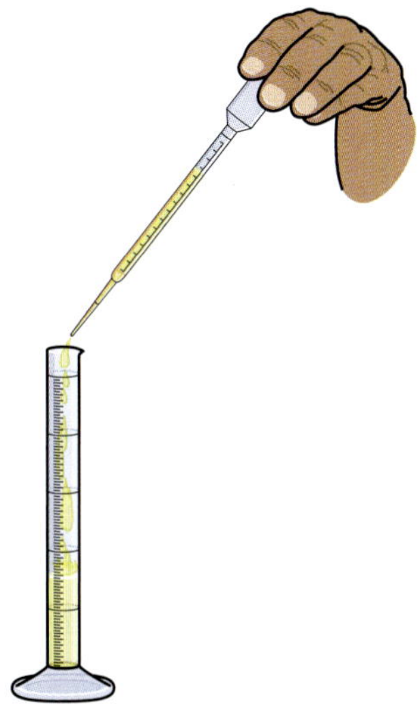

7. Weigh the empty 250-mL beaker, and record the mass to the nearest 0.01 g. Alternatively, if you are using a digital balance, place the beaker on the balance and select the "tare" option.

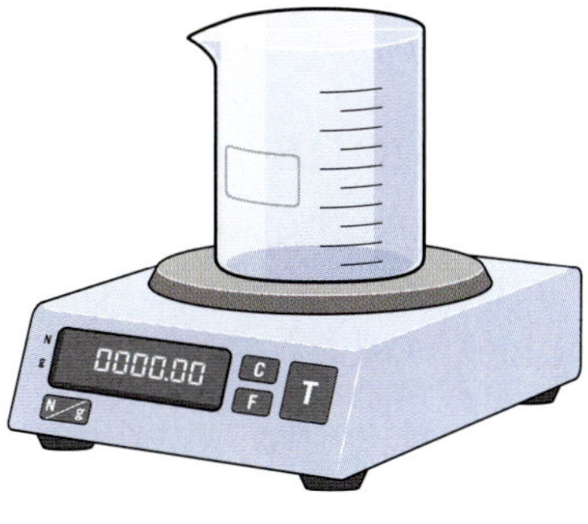

8. Completely fill a numbered pipette with the first sample liquid.

9. Invert the filled pipette and place it in the beaker. Determine the weight of the filled pipette to the nearest 0.01 g. Record the data on your table. (On non-digital balances, you will need to subtract the weight of the beaker to determine the weight of the filled pipette.)

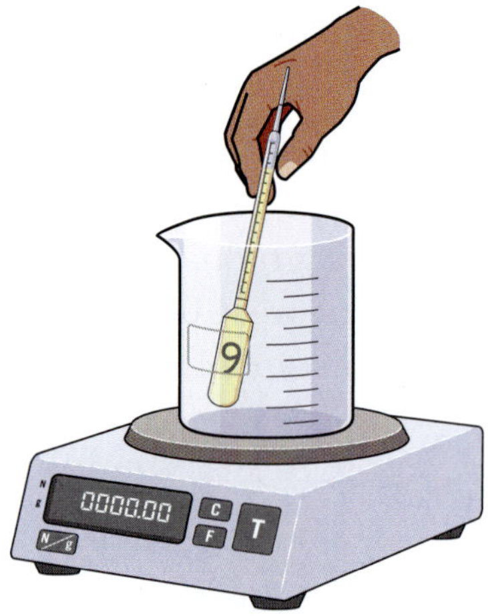

10. Repeat the procedure for each of the sample liquids and for the distilled water.

11. Save the filled pipettes for another procedure.

12. Find the density of each liquid sample by dividing the weight of the liquid in grams by the volume of the sample in mL. As Table 4 indicates, to find the weight of the liquid, you must subtract the weight of the pipette from the total weight of the filled pipette. Record the density of each liquid in grams per milliliter.

Analyzing the Data

- Which liquid has the lowest density? Rank the liquids in order from lowest density to highest density.

Get the names of the five test liquids from your teacher and find their densities in the *Handbook of Chemistry and Physics* (D. E. Lide, Ed.; CRC Press, Inc.; Boca Raton, FL 2000) or another chemistry handbook. You may need to compare your data to those of another team.

Arriving at Conclusions

Compare the densities of the five liquids to your data and use this information to identify each sample. Share your findings with other teams and with your teacher.

- How accurate were you in identifying the test liquids?
- How could you improve your accuracy?

Connecting to the Problem

Explain how the density test helps the technician identify the components of a mixture in a sample of crude oil. You will need to include this density test on your laboratory flowchart at the end of your technician training.

Laboratory 3 Another Method of Determining Density of Test Liquid

Overview

In this laboratory you will use a hydrometer to measure the specific gravity of the same five liquid samples that you tested in Laboratory 2.

Background

Finding the specific gravity of a liquid is another way to determine its density. **Specific gravity** is defined as the ratio of the density of a material to the density of water. Since in such calculations the units cancel one another, specific gravity is expressed as a unitless number. For example, the density of a common type of 10W-30 motor oil is 0.88 g/mL. When this is compared to the density of water at 3.98 °C (1.0 g/mL), the units (g/mL) cancel out, and the specific gravity of the motor oil is given as 0.88. You will find specific gravities for liquids given on most MSDSs. A hydrometer is an instrument designed to measure the density of a liquid in terms of its specific gravity. The instrument takes advantage of Archimedes' principle, which states that when a body is immersed in a fluid, the fluid exerts an upward force on the body equal to the weight of the fluid that is displaced by the body. The hydrometer is weighted to duplicate the density of water. Acting like a buoy, the hydrometer takes a position in the test liquid so that the point at which it breaks the surface can be read on a scale. The hydrometer provides a relatively easy and quick way to determine a liquid's density.

Materials

- safety goggles (1 pair per student)
- aprons (1 per student)
- 1 set of 5 liquids to be tested, 100 mL each, at room temperature
- 100 mL distilled water, at room temperature
- 1 marking pen
- 6 200-mL graduated cylinders

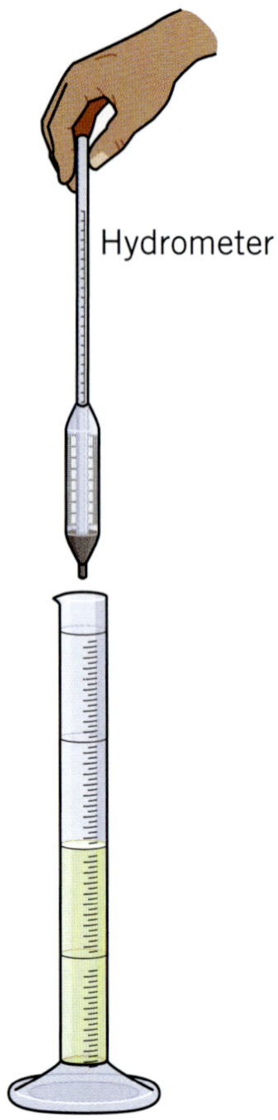

Figure 4
The scale within the hydrometer will indicate the specific gravity of the liquid.

- 1 hydrometer, universal scale
- MSDSs for test liquids
- S/H/E checklist

Safety

Using your S/H/E checklist, prepare a S/H/E plan for this procedure. Be sure it includes the following guidelines:

- Wear safety goggles and a laboratory apron throughout the procedure.
- Remember, tasting is never appropriate in the chemistry laboratory.
- Be careful when handling glassware.
- Dispose of used solutions as your teacher directs.

Procedure (for 3 or 4 students)

1. Read through the entire laboratory procedure, and read the MSDSs for the test liquids. (The MSDS sheets will be coded to the numbers of the test liquids but will not have the names of the liquids on them.) Using your checklist, make a S/H/E plan for this laboratory test. Make sure that you understand what you need to do and what precautions you need to take.

2. Pour 100 mL of each numbered test liquid into the 200-mL graduated cylinders. Mark each cylinder to identify the contents.

3. Using the data from the previous activity as a guide, rank the liquids from lowest density to highest density.

4. Place the hydrometer in the first cylinder; record the specific gravity by aligning the surface of the liquid to the point on the scale closest to that surface (Figure 4).

Analyzing the Data

- How did the specific gravity readings for each liquid compare to the density calculations?
- How might you explain any differences?

Arriving at Conclusions

- Compare your specific gravity readings to those provided on the MSDSs. What are the identities of the test liquids?

If there is a great difference between your reading of specific gravity and the information on the MSDS, you may need to repeat the test. You can also compare your team's results to those of other teams.

- Which procedure (Laboratory 2 or Laboratory 3) was most useful in identifying the liquids? Why? note

Connecting to the Problem

Part of the role of the laboratory technician in the petroleum refining process is to analyze crude oil samples according to their components. Think about how specific gravity could be used to identify components in crude oil. Would specific gravity be useful in identifying the mixture as a whole?

Later you will include the standard test for specific gravity on your flow diagram for the petroleum refinery laboratory.

Laboratory 4 Standard Test for Pour Point

Overview

In this laboratory you will determine the **pour point** of a sample by observing the lowest temperature at which a sample is able to flow.

Background

Pour point is an important physical property since it predicts how the product will perform at various temperatures. At low temperatures, a liquid petroleum product might solidify or form a gel or a crystallized wax. Usually, products intended for use in cold environments must have lower pour points than products designed for warmer climates.

In this test, first a sample is heated; then it is slowly cooled in a water bath. The sample's characteristics are examined after each 3 °C change in temperature. The technician tilts the test cylinder slightly to see if the sample shows any movement. Finally, the temperature is lowered to the point below which the sample does not move when tilted. The temperature is the sample's pour point.

Materials

- safety goggles, gloves, and laboratory apron
- 1 flat-bottomed cloud-and-pour-point jar, approximately 30 mm X 120 mm (For an alternative you can use a graduated cylinder removed from its detachable base.)
- 1 alcohol thermometer calibrated to the nearest 1 °C and capable of measuring temperatures of –35 °C or lower
- 1 one-hole cork (not rubber) stopper that fits the cloud-and-pour-point jar and can hold the thermometer firmly in place
- 1 tube stopcock grease (all groups will share this)
- 1 copper beaker with diameter and depth great enough to hold the cloud-and-pour-point jar as shown in the diagram

- 1 cork washer to hold the cloud-and-pour-point jar in middle of copper beaker
- 1 large (1000–2000 mL) beaker to hold the dry ice/isopropanol mixture "bath" and the copper beaker
- 1 cork disk to place under copper beaker for insulation, cut from a sheet of thin cork
- 1 box large enough to hold the beaker—you will also use this as a mold for Styrofoam insulation
- 1 can of "spray-in" Styrofoam solution to make insulation
- 1 or more paper towels
- 1 glass marking pen, permanent
- 1 ring stand
- 1 stirring device (spoon, tongue depressor, or stirring rod)
- supply of dry ice in 3 cm x 3 cm pieces, as needed
- 99% isopropanol as needed (Note: 70% isopropanol will freeze at these temperatures)
- 1 test sample supplied by your teacher
- MSDS sheets
- S/H/E checklist

Test Design Challenges

Your team will first need to construct a "Dewars" liquid bath setup, which will provide a safe method for holding an isopropyl alcohol (isopropanol) and dry ice mixture. The container for the isopropanol and dry ice should be a standard laboratory glass beaker large enough to submerge a flat-bottomed copper vessel to a depth at least equal to that of the oil sample within the enclosed jar. The beaker must also be wide enough in diameter to allow adequate room for the isopropanol-dry ice mixture to be stirred during the test (Figure 5). A Dewars setup enables you to measure temperatures as low as –35 °C. The following information will help you to carry out this challenge:

1. First prepare an insulated container to hold the large glass beaker. Choose a cardboard box that is both taller and wider than the beaker by 2–3 inches in all directions. Spray liquid Styrofoam on the bottom of the box to a depth of about 2–3 inches. When this layer dries, set the beaker in place, and spray foam around the sides of the beaker to form a layer of surrounding insulation. You can leave the beaker in place or carefully remove it when the foam dries. The insulation will form a well for holding the beaker.

2. Mix dry ice pieces with 99% isopropanol. It will take about 10 minutes of stirring the isopropanol mixture to achieve the desired

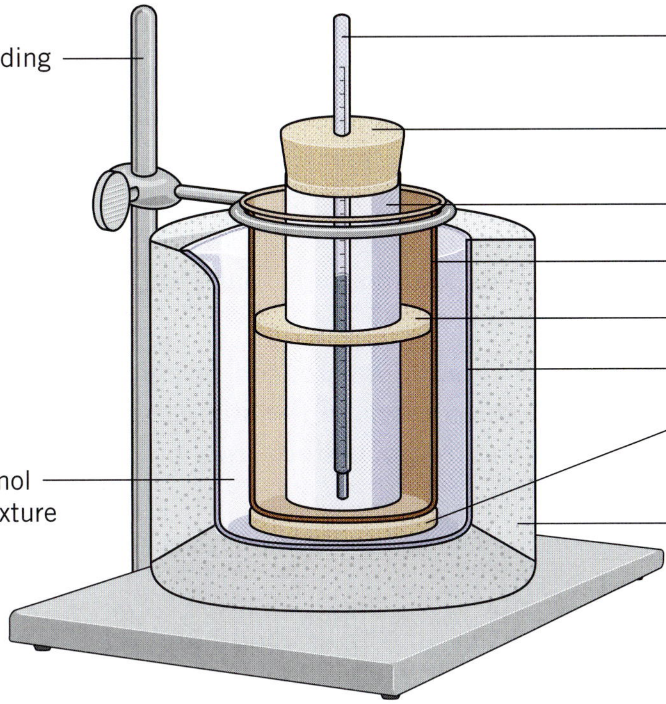

Figure 5
A cutaway of a completed Dewar's setup

temperature—as close to –35 °C as possible. Note: It is necessary to swirl the pieces around to break off any water ice that forms on the surface of the dry ice pieces. The water ice will insulate the dry ice from the isopropanol. Never touch dry ice.

3. Now, assemble the Dewars set-up as shown in the diagram.

Safety

Using your S/H/E checklist, prepare a S/H/E plan for this procedure. Be sure it includes the following guidelines:

- Wear safety goggles and a laboratory apron throughout the procedure.
- Remember, tasting is never appropriate in the chemistry laboratory.
- Be careful when handling glassware.
- Know the safe procedure for inserting a thermometer into a cork to avoid accidental breakage. Insert the thermometer according to your teacher's directions.
- Avoid skin contact with any liquids or solids.
- Exercise caution when working with cold solutions. Do not touch them with your hands. Work with dry ice only under an operating fume hood. Wear heavy gloves and use tongs when handling dry ice.
- Read over the MSDS sheets provided for the substances your class will be using in the laboratory.
- Dispose of used solutions as your teacher directs.

Procedure (for 3 or 4 students)

1. Read through the entire laboratory and the MSDS for the test liquid that is assigned to your group. Each group will get a different test liquid. Your MSDS sheet will be coded to the number of your assigned test liquid but it will not have the name of the liquid on it. Make a S/H/E plan, using your checklist. Make sure that you understand what you need to do and what precautions you need to take for the entire procedure.

2. Apply a small amount of stopcock grease to the thermometer tip. Holding the thermometer and the stopper with a heavy cloth towel, carefully insert the thermometer through the cork, as demonstrated by your teacher.

3. Assemble the Dewars equipment described in the earlier design challenge (p. 37).

4. Measure the length of your cloud-and-pour-point jar, and make a mark at the halfway point. Then make a second mark below it at the one-quarter length of the jar.

5. Test the position of the thermometer in the cork by inserting the cork in the cloud-and-pour-point jar. Adjust the thermometer so that the tip will be suspended at the one-quarter height mark of the jar. Remove the cork.

6. Obtain your assigned room-temperature sample liquid, and pour it into the cloud-and-pour-point jar to the halfway mark.

7. Place the cork with the thermometer firmly into the cloud-and-pour-point jar.

8. Set up the ring stand to hold the copper flask. Insert the cloud-and-pour-point jar into the copper beaker using the cork washer to ensure the jar remains vertical inside the beaker.

9. Fill the large glass beaker about 1/3 full with the dry ice-isopropanol mixture. Set it in its insulated container.

10. Using the ring stand clamp, carefully lower the copper beaker assembly into the liquid bath. Stir this outer bath frequently during the procedure.

11. Note the initial temperature of your test sample.

12. Allow temperature to drop 3 °C. Pull the jar out and tilt it slightly. If the liquid moves, replace the jar, vertically. Do not allow the jar to remain out of the copper beaker for more than a few seconds.

13. Repeat step 12 for each 3 °C drop in temperature until the liquid no longer moves when the tube is tilted.

14. When the sample no longer shows any movement, hold it in a horizontal position for 5 seconds. If the sample still shows no movement, it has exceeded the pour point.

15. Record the pour point as the stop temperature minus 3 °C. This is the lowest temperature at which the sample can be poured.

Analyzing the Data

- Each team received a different sample to analyze. Rank the samples from the highest to the lowest pour point temperatures.

Look at the MSDE sheets for the samples. What relationships do you see between the pour point temperatures and other chemical and physical properties of the samples?

Arriving at Conclusions

If the liquids sampled by your class represent refined petroleum products, suggest possible uses for these products.

- Which would be suitable for cold climates or for refrigerated environments?
- Which might perform best as lubricants for heated metal machinery parts?

Connecting to the Problem

The pour point of a petroleum product affects its uses because it affects how the product performs under working conditions. The pour point also affects how crude moves through transporting pipelines from drilling sites. It also affects its flow within the refinery as it undergoes various processes and experiences different temperatures.

At the end of your technician training, you will include the pour point test on your flow diagram of laboratory tests.

Laboratory 5 Test for the Percent Volume of Water in a Mixture

Overview

In this laboratory you will use a special centrifuge tube etched with volume markings to measure the volume of water contained in a sample mixture. You will use this volume to calculate the percent water content of the sample.

Background

Crude oil is a mixture of liquids, each of which varies in its physical and chemical properties. Solubility is a physical property of particular importance to oil refining. The technician uses varying solubilities to separate certain components of the mixture. You may notice this property when performing a practical kitchen task—making salad dressing. Combine water, vinegar, and oil, and you soon find that the old saying is true: "Oil and water don't mix." The vinegar and the water will combine, but when oil and water are stirred together, they quickly separate to form two distinct layers of liquid.

Work at the refinery begins when samples of crude oil containing various amounts of water and dissolved solids arrive. The technician is often assigned the initial task of finding out the relative amounts of these materials in the crude's total mass. The first step of the analysis involves placing a sample of the raw crude oil in a centrifuge tube and spinning at high speed (Figure 6). Substances of greater density sink toward the bottom of the tube; substances of lesser density rise to the top.

In this laboratory you will perform a centrifuge test. When the tube stops spinning, you will examine the sample. Any water in your sample will collect at the bottom of the tube. This layer will also include any water-soluble components in the original mixture. In the refinery, technicians perform this test to analyze all crude oil scheduled for refining. Samples containing too much water or other sediments may go through extra treatments prior to processing.

Figure 6
A centrifuge for separating components by density. Note that the test bottles must be set so that they balance in the centrifuge.

Materials

- safety goggles (1 pair per student)
- apron (1 per student)
- 2 mL of simulated crude oil sample assigned to your group
- 2 mL deionized water
- 1 10-mL graduated cylinder
- 2 centrifuge tubes calibrated in mL
- 1 stopper to fit centrifuge tube
- centrifuge to be shared with other groups
- MSDSs for test liquids
- S/H/E checklist

Safety

Consult the S/H/E plan you made in your safety training and the information below:

- Wear safety goggles and a laboratory apron throughout the procedure.
- Tie back long hair and secure loose sleeves when working around the centrifuge.
- Exercise caution when using the centrifuge. Use it as demonstrated by your teacher. Always monitor a running centrifuge because its vibration can cause it to "walk" off a counter top. Do not try to slow the moving centrifuge with your hands, and remove tubes only when the centrifuge has come to a complete stop.
- Do not allow any chemicals to come in contact with your skin. In case of accidental contact, wash with soap and water.
- Dispose of chemicals at the end of the lab as directed by the teacher.

Procedure (for 3 or 4 students)

1. Read through the entire laboratory and the MSDS for the material you are using in the laboratory. Use your S/H/E checklist to make a S/H/E plan for this laboratory. Make sure that you understand all directions and precautions before you begin the test.

2. Pour 2 mL of sample into a centrifuge tube. In your laboratory notebook write a description of the physical appearance of the sample. Note color, clarity, viscosity, presence of suspended material, and any other observable features of the liquid.

3. Add 2 mL of water to the centrifuge tube.

4. Stopper the tube, and while holding the stopper in place, invert it a minimum of 10 times until the liquids are mixed. In your laboratory notebook note any changes in appearance of the liquid.

Figure 7
Place the sample tube in one of the ports in the center "nest" area of the centrifuge. Then fill another tube with water and position it in the port opposite the sample tube. This will provide balance as the centrifuge spins.

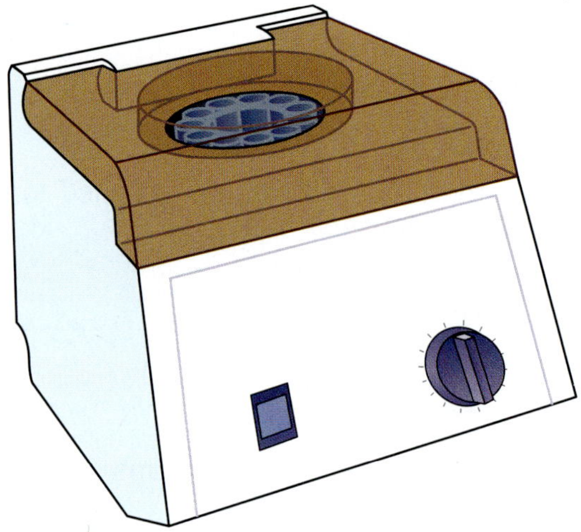

5. Remove the stopper from the centrifuge tube, and place the tube into a centrifuge port as directed.

6. Pour 4 mL of water into the second centrifuge tube, and place it in the port opposite the one occupied by your sample tube. This will provide a balance for the centrifuge apparatus (Figure 7).

7. Spin for a minimum of 5 minutes at a relative centrifugal force (rcf) of 500.

8. After the centrifuge comes to a complete stop, remove the sample tube and examine the contents. Describe its physical appearance in your notebook.

9. Record the volume of the water and sediment present in the bottom of the tube in this first trial.

10. Repeat steps 3–9. Again, record the volume of water and sediment present in the bottom of the tube. Continue to repeat these steps until you record two identical volumes.

11. Without disturbing the contents of the tube, examine the physical appearance of the two layers. How do they compare to the appearance of the original sample?

12. Do not remix the layers in the tube. You will need them both for another procedure.

Multimedia Link

CD-ROM In the Technician Tasks section of the CD-ROM, look up the method for calculating the percent of water in a sample by using a centrifuge.

Analyzing the Data

Use your final volume data to calculate the percentage by volume of water and sediment in your sample. Remember that you are considering the water that appeared in the bottom of the tube *in addition* to the water that you added to the tube. Your percentage should express the relationship of the additional water-sediment volume to the total volume of your original sample. Post your percentage calculation on a chart in the front of the room. note

Arriving at Conclusions

- Examine the water and sediment layer on the bottom of the tube after centrifuging. In which physical property does this layer differ from the top layer in the tube?

In a refining facility, if the percentage of water and sediment in the crude sample is greater than 1–2%, the crude will require additional treatment before it enters the refining process. Why would the additional treatment be important?

- Which of the samples of simulated crude analyzed in your class will be suitable for refining without further treatment?

Connecting to the Problem

Raw or untreated crude oil may contain sediments of dirt and rocks that were acquired during the drilling process. In addition to this visible material, water residues in crude oil may contain other materials in solution. The technician tests the aqueous layer for metals since metallic residues in the crude would affect the quality of any refined petroleum products. You will perform such a test in the next laboratory.

Laboratory 6 Test for the Presence of Dissolved Solids in a Mixture

Overview

In this laboratory you will establish a table of color standards for various metals in solution by performing flame tests. Then you will test the aqueous layer from your crude sample from Laboratory 5 to identify any metal(s) in solution.

Background

In this laboratory you will test the solution at the bottom of the centrifuge tube from Laboratory 5 for the presence of metal ions. To do this, you will expose a small amount of solution to the hottest part of a burner flame. The flame will vaporize the water, then quickly vaporize the salts left behind. The intense heat will cause the ions to dissociate into atoms. The flame excites the electrons surrounding these atoms to a higher energy state. When the electrons return to a lower energy state called **ground state,** they emit energy in the form of light. The wavelength pattern, or color of this light, is unique for each metal, so a visible color will be your clue for identifying the metals in solution.

Materials

- safety goggles (1 pair per student)
- aprons (1 per student)
- 1 set of 6 standard solutions:
  - ➤ Calcium chloride
  - ➤ Nickel(II) nitrate
  - ➤ Copper(II) nitrate
  - ➤ Lithium nitrate
  - ➤ Strontium nitrate
  - ➤ Barium nitrate
- distilled water
- centrifuged sample from Laboratory 5
- 1 disposable pipette
- 1 nichrome wire in a holder
- laboratory gas burner and striker
- MSDS sheets
- S/H/E checklist

Safety

Consult the S/H/E plan you made in your safety training and the information below:

- Wear safety goggles and a laboratory apron throughout the procedure.
- Tie back long hair and secure loose sleeves when working around a flame.
- Clear the table top of any flammable materials.
- Light the gas burner as directed by your teacher.
- Do not touch nichrome wire after heating.
- Dispose of solutions as your teacher directs.

Procedure (for 3 or 4 students)

1. Read through the entire laboratory and the MSDSs for the materials you are using in the laboratory. Use your S/H/E checklist to make a S/H/E plan for this test. Make sure that you understand all directions and precautions before you begin work.

2. Set up a table in your laboratory notebook for recording the flame colors for each solution in the standard set.

Science Connections
Desalting Crude Oil

Salts must be removed from crude oil before the refining process begins. This is because these salts can corrode the refining equipment. Desalting and dehydrating crude oil is generally accomplished by chemical and electrostatic methods involving the use of hot water as an extracting agent. Another method involves filtering the crude oil.

Chemical desalting uses a chemical called a **surfactant,** which, when added to water, causes water molecules to become less attracted to one another, thus making water a more effective solvent for other molecules. Detergent is an everyday example of a surfactant. It allows water molecules to surround dirt and grease particles for removal from fabrics and surfaces.

In chemical desalting, the crude oil is placed in a settling tank. As the added hot water settles to the bottom, impurities are extracted from the crude. Surfactants are only added when technicians, analyzing an incoming crude, report that it has a particularly large load of suspended solids. In electrical desalting, high voltage is applied to concentrate the suspended water globules quickly in the bottom of the settling tank.

In a third method, heated crude is filtered through a layer of diatomaceous earth, a soil rich in fossil remains of diatoms. This highly porous material is often used as a filtering agent. Suspended solids in the sample stick to this material as the crude oil passes through. The desalted and dehydrated crude is ready to enter the refining tower.

---

3. Light the gas burner with a striker as directed by the teacher. Adjust the air vent on the burner until all traces of yellow or orange disappear from the flame.

4. Hold the nichrome wire by the attached holder. Place the wire in the flame until it glows red. Allow it to cool slightly.

5. Dip the wire into the first solution on your standard list. Place the wet wire into the hottest part of the burner flame (Figure 8) and observe the resulting flame color. Record the solution and the color.

6. Dip the wire into distilled water.

7. Repeat steps 5–7 until you have tested all of the standard solutions.

8. Using the disposable pipette, draw off the top layer of liquid from your centrifuged sample and discard the liquid as directed by the teacher.

9. Dip the wire into the remaining layer, and flame test the solution in the same way you tested the standards. Record the color of the flame.

10. Dip the wire into distilled water; then reheat the wire. Allow the wire to cool before placing it on the table.

Figure 8
Position the wire in the luminous part of the flame, not in the center cone.

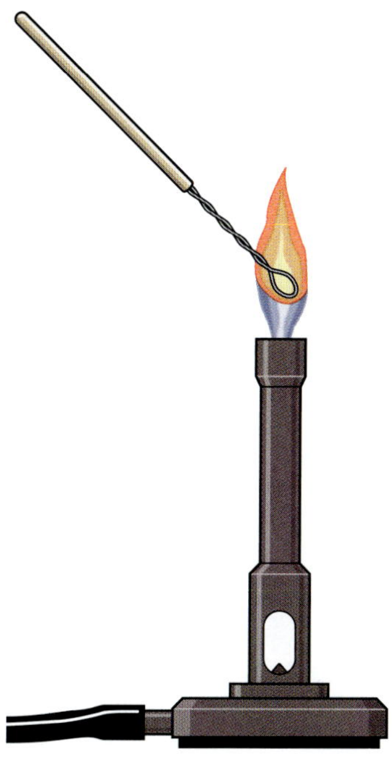

Analyzing the Data

Compare the flame test result for your sample to your table of standard tests.

Arriving at Conclusions

- What metal ion or ions are probably in your sample? What evidence do you have for your answer?

Compare your findings with those of the other groups.

- Are all samples the same? In what ways do they differ?

Connecting to the Problem

The water present in the crude and the solids dissolved in that water are contaminants that must be removed before refining begins. Otherwise their presence would cause corrosion, plugging, and fouling of equipment as well as the poisoning of catalysts used in the process.

When you have completed your technician training, you will include the test for the presence of dissolved solids on your flow diagram for the oil refinery laboratory.

Laboratory 7 Corrosion Test for Detecting Sulfur

Overview

In this laboratory you will analyze samples of simulated crude oil to determine the relative amounts of sulfur compounds that are present.

Background

In the refining process, sulfur is removed from petroleum. Sulfur compounds in petroleum products can have a corrosive effect on metals. The copper corrosion test (Figure 9) analyzes gasolines, solvents, fuel oils, lubricating oils, and other petroleum products for sulfur.

Figure 9
Stages in the corrosion test

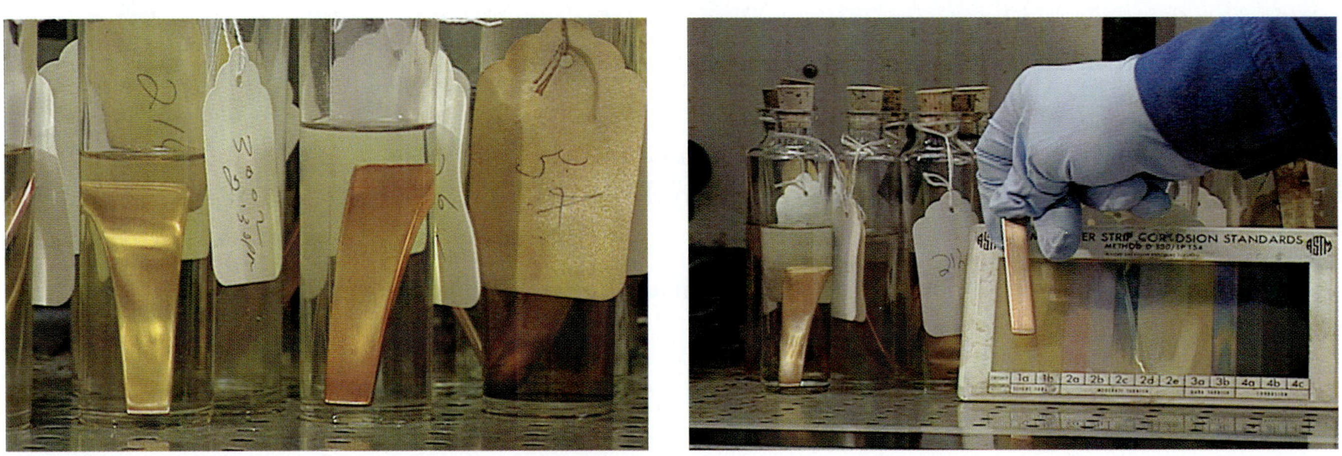

Materials

- safety goggles (1 pair per student)
- aprons (1 per student)
- 4 test tubes, 25 x 150 mm
- 4 corks to fit tubes
- thermometer for monitoring water bath or incubator
- 4 pure copper metal strips, cut to fit within the stoppered test tubes
- duct tape
- 1 square each of 240-grit and fine sandpaper
- piece of cotton cloth
- permanent marking pen, fine tipped
- 4 metal forceps for removing strips
- 4 250-mL beakers
- 4 covered glass or plastic vials large enough to hold strips
- MSDSs for test materials
- wash bottle containing acetone
- absorbent towels
- S/H/E checklist

Materials for entire class

- disposable gloves, safety goggles, and laboratory apron
- 4 test samples of simulated crude in bottles marked A, B, C, and D
- bacterial incubator set to 50 °C or a hot water bath, deep enough to submerge the entire tube and capable of maintaining a steady 50 °C ±1 °C temperature for 3 hours

Safety

Consult your S/H/E checklist and the information below:

- Wear laboratory coat or apron, goggles, and protective gloves when handling samples.
- Acetone is very flammable and must be kept away from any flame or sparking equipment. Use acetone solvent as directed under a well-ventilated hood.
- Handle copper strips with care to avoid cuts.
- Some samples may be flammable. Keep them away from open flames and electrical sparks.
- Dispose of samples as your teacher directs.

Procedure (for 3 or 4 students)

1. Read through the entire laboratory and the MSDSs for the materials you are using in the laboratory. Using your S/H/E checklist, make a

S/H/E plan for this test. Make sure that you understand the procedures and precautions before you begin this test.

2. Put a piece of duct tape over the narrow edge of each of your copper strips. Handle the strips by holding onto the duct tape, not the edges of the copper.

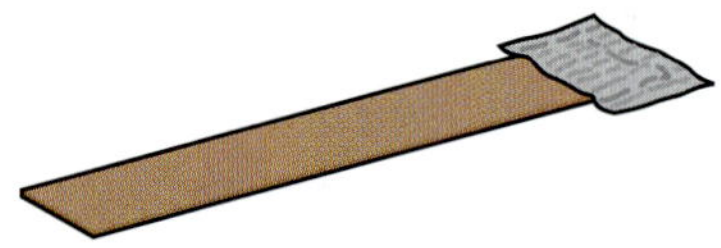

3. Polish both sides of the copper strips first with heavy, then fine sandpaper to remove any traces of oxidation. Then wipe the strips with cotton cloth.

4. Handling strips carefully by the duct tape strip, place one strip in each test tube. Then remove the tape.

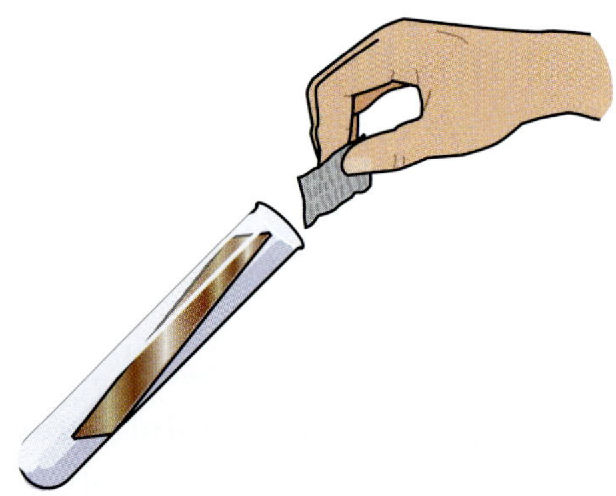

5. Use the marking pen to mark tubes with the letters A, B, C, and D.

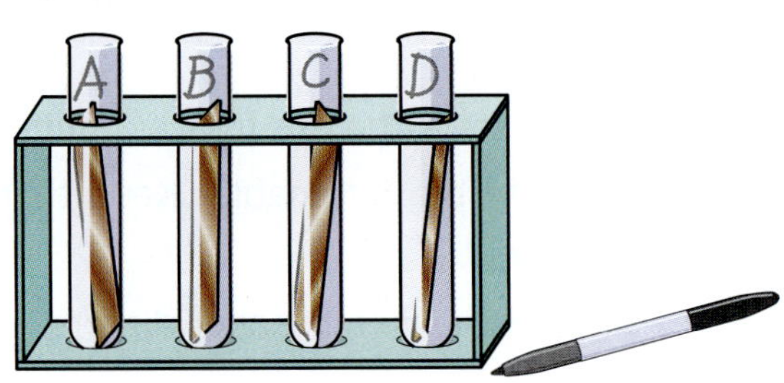

6. Carefully pour test sample A into tube A until the copper strip is submerged but do not fill the tube to the very top. Leave enough head space to place the cork in the tube. Repeat this procedure for tubes B, C, and D.

7. Place a cork firmly in each tube.

8. Mark each tube with an identifying symbol for your group.

9. Place the tubes in the incubator or the water bath as directed by your teacher.

10. Wait 3 hours or overnight.

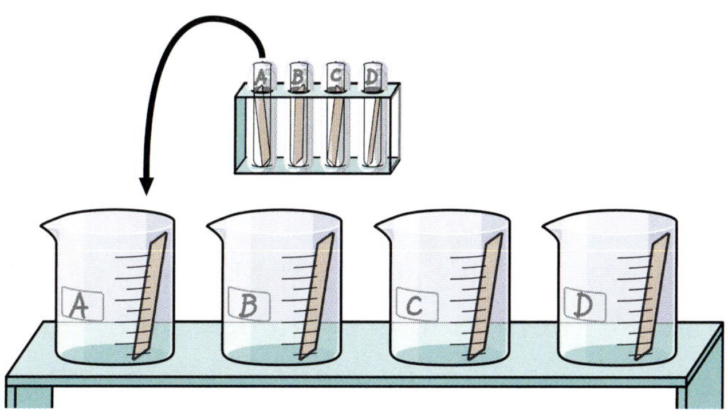

11. Remove tubes from the heat source.

12. Mark four beakers with the letters A, B, C, and D.

13. Using the forceps, transfer the strip from tube A to Beaker A. Repeat the procedure for strips B, C, and D.

14. Working under a ventilated hood and wearing protective gloves, carefully wash the surfaces of each strip with the acetone solvent to remove all traces of the sample solutions. Do not allow acetone to come in contact with your skin. Blot each surface gently with a piece of absorbent towel. Do not rub. *NOTE:* Acetone is highly flammable. Keep it away from open flames or electrical equipment.

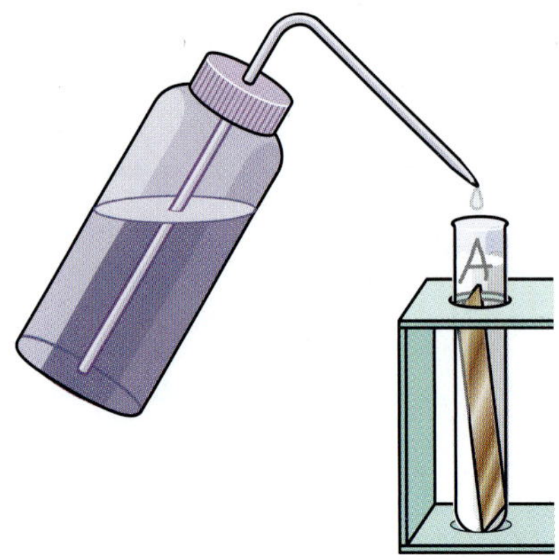

15. If you will not be able to analyze the strips immediately, transfer them to vials marked A, B, C, and D. Identify the vials with your group symbol, and apply covers to avoid exposure to air. Do not place the vials in a strong light.

Multimedia Link

CD-ROM

You can practice your skill with the copper corrosion test on the Technicians Tasks and Practice sections of the CD-ROM.

Analyzing the Data

Examine the four copper strips for the appearance of bluish copper sulfate ($CuSO_4$), or tarnish, on the surface. Rank the four samples from least tarnished to most tarnished.

Arriving at Conclusions

Your teacher will tell you the chemical identities of the four test solutions. Which of the solutions contained sulfur compounds at the greatest concentrations? The least? Were the results of your copper corrosion tests what you would have expected for these solutions? Explain.

Connecting to the Problem

Sulfur compounds are undesirable in petroleum products. They contribute to corrosion and have bad smells. When the petroleum product is combusted, sulfur compounds contribute to air pollution. It is important that technicians can estimate how much sulfur is in a sample, so that plans can be made to remove as much as possible.

At the end of your technician training, you will place this corrosion test for detecting sulfur on the flow diagram for the petroleum refinery laboratory.

Laboratory 8 Testing for Clarity of Samples

Overview

In this laboratory you will use the CD-ROM to learn how technicians test samples of crude oil for their clarity.

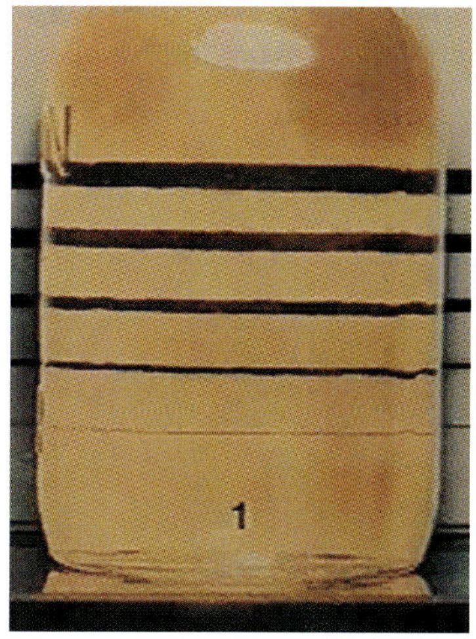

Background

Petroleum refinery technicians use the clarity of a sample as an indicator of the purity of the sample. Sediments from the extraction process may remain in a sample, making it look cloudy.

Materials

- disposable gloves, goggles, and laboratory apron
- CD-ROM and player
- laboratory notebook and pen
- solutions from Laboratory 7: Corrosion Test
- safety goggles (1 pair per student)
- aprons (1 per student)
- plastic gloves
- 5 250-mL beakers, 1 for each sample plus 1 for distilled water
- white poster board
- black markers
- safety scissors
- distilled water
- wax marker to label beakers
- S/H/E checklist

Safety

Consult your S/H/E checklist and the information below:

- Wear safety goggles, protective gloves, and a laboratory apron throughout the procedure.
- Be careful when handling glassware.
- Dispose of samples as your teacher directs.

Procedure (for 3 or 4 students)

1. Read through the entire laboratory procedure, and review the MSDSs for the test liquids saved from Laboratory 7. Using your S/H/E checklist, make a S/H/E plan for this laboratory test. Make

sure that you understand what you need to do and what precautions you need to take.

2. Access the CD-ROM Technician Tasks section, and click on the test for Clarity Analysis. This part of the CD-ROM will help you to learn the procedure for the test.

3. Once you feel confident that you understand the procedure, go to Practice on the CD-ROM, and try your skill at determining the clarity of two samples. Keep practicing until you can make an accurate judgment of a sample's clarity.

4. The CD-ROM provides an example of a tool that technicians use to evaluate clarity, a card with horizontal lines graduated in thickness. In the laboratory, using the example in the CD-ROM as a model, make up five such cards.

5. Label each 250-mL beaker with the names of one of the samples or with the label "distilled water." Pour equal amounts of the liquids into the beakers.

6. The distilled water is your standard. Hold the beaker of distilled water in front of one of the cards, and look through the beaker to check for clarity. This should be considered a clarity reading of 1.

7. Range your other samples alongside the distilled water beaker, each with a lined card behind it (Figure 10). Move the beakers and their cards around until you have ranked the samples from the most clear to the least clear. Assign a number ranking to each.

8. Compare your team's clarity levels with those of another team. Try to reach an agreement on the ranking.

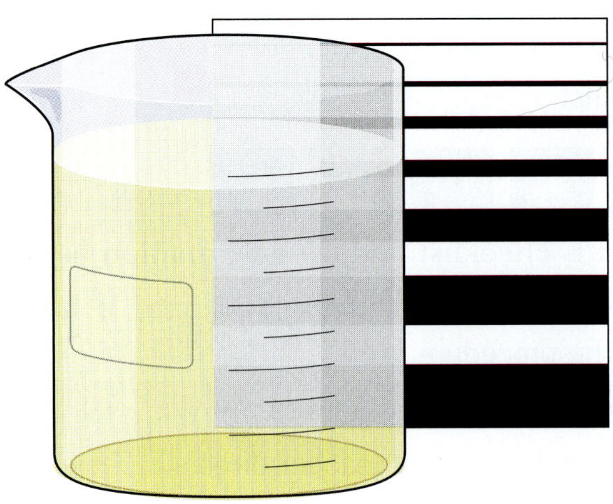

Figure 10
Using a card to determine clarity

Analyzing the Data

Look over all your samples and their clarity rankings.

- How did the clarity test help you determine the "purity" of your samples?
- What reduced the clarity of the sample that was least clear?

Arriving at Conclusions

If you read the background section to this laboratory, you will remember that sediments in crude oil contribute to a lack of clarity. Based on your own laboratory experience, how valid a test is the card test for determining the presence of sediment in a sample?

Connecting to the Problem

Sediment in a sample of crude oil must be removed before the crude can receive further treatment. One of the jobs of the laboratory technician is to determine the extent to which the sediment removal will be necessary for a sample. How well did the CD-ROM help you learn this skill? How could you use the CD-ROM to assess your ability to perform this skill?

You will now be asked to include this test on the flow diagram for the oil refinery laboratory.

data and results

This training program was designed to help you, as a trainee technician, become proficient in a set of science concepts, laboratory skills, and workplace skills. The focus problem for the module was this: How can a petroleum refining company test incoming samples of crude oil?

Developing a Flow Diagram for Testing a Sample of Crude Oil

Recall that a thorough and accurate laboratory analysis on a sample of crude oil is essential so that company officials can effectively evaluate the material before purchasing a large quantity. If the company is prospecting for new sources, the analysis of a sample of crude oil will assist officials in deciding whether to invest further resources in developing a particular oil field. Even after the crude is accepted, the laboratory analysis forms the basis for planning the refining steps needed to yield the desired products. Clearly, your work as an oil refinery technician is essential to the overall success of the industry.

To ensure accurate, safe, and consistent analyses for every incoming sample, you will work with your colleagues to develop a flow diagram for each of the testing procedures carried out in your laboratory. This graphic display of the procedures and protocols will help to standardize the tests for reliable and consistent reporting of data. By including appropriate safety checklists, you will reinforce the S/H/E procedures mandated for each test.

Study examples

Begin by examining examples of industry flow diagrams. A good source of diagrams is the Technical Manual of the Occupational Safety and Health Administration (OSHA) of the U.S. Department of Labor. This manual, available on the OSHA Website *http://www.osha-slc.gov/dts/osta/otm/otm_iv/otm_iv_2.html* displays a complex process flow diagram detailing all of the steps involved in turning crude oil into useful products. For each process listed on the refinery diagram, there is a separate flow diagram displaying the key steps in order. One of these flow diagrams (Figure 11, page 56) shows the electrostatic process for desalting incoming crude oil, an initial step in refining:

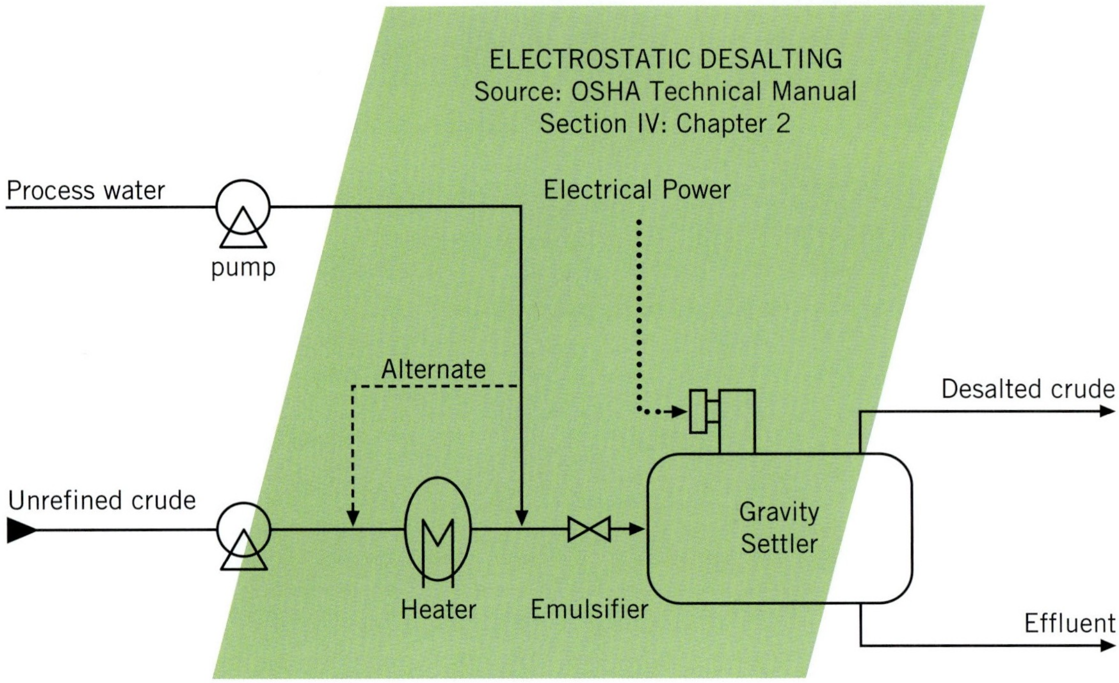

Figure 11
An industry process flow diagram

Present information

Notice that the process flow diagram effectively and efficiently displays a great deal of information. Although you have not been trained to operate the desalting equipment, you can use the flow diagram to answer several questions:

- What are the materials needed to perform the electrostatic desalting of crude?
- What is the energy source?
- Is the crude oil heated before or after it is exposed to electrical energy?
- At what point are water and crude oil mixed? Is there more than one way the two can be mixed?
- What are the end products of electrostatic desalting?
- After this process is complete, what do you think will happen to the end products?

Develop icons

Notice that small drawings are used to represent pieces of equipment. You may be thinking that it would be just as easy to simply write "heater" without taking the time to draw the picture. That would be true if this were the only flow diagram that included this item. However, this drawing can now symbolize "heater" in future diagrams. The drawing becomes an icon for rapid visual communication of the equipment and its function.

Think about icons that you encounter in your daily life. Examples are stop signs, restroom signs, on-off symbols, and more. Make a list of at least five icons that you have encountered in the last 5 hours.

This *Science in a Technical World* module also displays icons to stand for information. Page through your student manual and note several of these. Be sure that you can decode the meaning of each icon you encounter in the module.

Present tests

Now begin the task of making flow diagrams for the various technician tests that you learned in your training. Your teacher will assign one of these tests to each group for initial planning. Working with the members of your group, sketch out a diagram that shows the steps of your assigned test. As you work, make a list of the steps and pieces of equipment that can be symbolized with icons. Finally consider your S/H/E plan as part of the diagram. *Be sure to include safety icons,* many of which are found throughout the pages of this STW module on your final flow diagrams. note

Develop lists

Share these lists with the entire class. Identify the steps and equipment mentioned on more than one list. For these items, design icons which everyone agrees to adopt. You may want to use computer drawing tools or software clip-art packages. Icons generated in this manner can be copied and pasted into all documents. Remember to keep the shapes and drawings very simple. You might want to consult manuals and online sources for ideas in designing these icons. If your class has contact with an industry representative, find out what symbols are adopted by technicians in the laboratory. note

Convert directions to flow diagram

Now convert the directions for each test to a flow diagram. Divide the work so that every group completes at least one diagram.

- Standard Distillation Test
- Standard Test for Density
- Another Way of Measuring Density of Test Liquids
- Standard Test for Pour Point
- Test for the Percent Volume of Water in a Mixture
- Test for the Presence of Dissolved Solids in a Mixture
- Corrosion Test for Detecting Sulfur
- Testing for Clarity of Samples

Before displaying these diagrams in their final form, groups should exchange drafts of the diagrams for review by at least one other group. In this way, groups can check for accuracy and clarity.

Your class should agree on a format for displaying the test diagrams in final form. Options include posters, pages in a notebook, power point slides, and overhead projector transparencies.

Linking the Technician Tasks

Each of the diagrams that you have completed will guide the technicians in performing the laboratory test accurately, safely, and efficiently. However, the analysis of an incoming sample of crude oil involves a set of tests, some of which must be performed in sequence. For example, recall that you saved the extracted water from your Test for Percent Volume of Water to be tested for dissolved solids in the next test. Clearly these tests are most efficiently performed in sequence.

Before planning a flow diagram for directing the sequence of tests needed to analyze an incoming sample of crude, look at another example of an industry process flow diagram.

In the flow diagram in Figure 12, you can see how individual processes become linked into a larger sequence. In the previous diagram, you saw the steps involved in desalting the crude oil. Are all of those steps outlined this time? What is included in place of all of

Figure 12

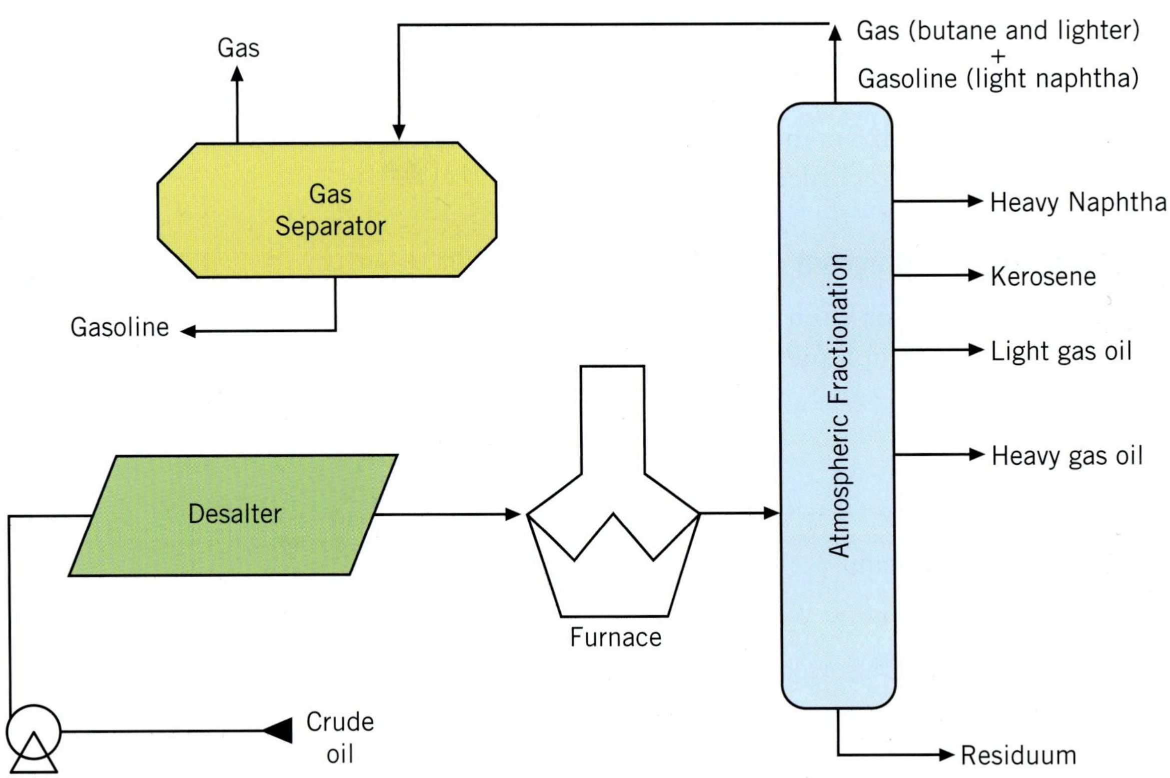

those individual steps? It's clear that there are two other processes in the set that are represented by symbols on this diagram—atmospheric fractionation and gas separation.

The diagram in Figure 12 is an industry *process flow diagram.* A true process flow diagram shows materials entering the process and products leaving the process. What are the materials and products shown above?

Now make a similar summary flow diagram for all of the tests on your laboratory list. Adopt a symbol for each of the tests in your set. Link the tests in the sequence you and your class agree is the safest, most effective, and most efficient way to test an incoming sample of crude oil. Display this laboratory flow diagram as the first page or the first poster in your laboratory sequence.

Evaluation Criteria

Both individual test and summary flow diagrams will be judged according to the following criteria:

- Were the individual steps in the processes sequenced correctly?
- Did the summary flow diagram illustrate logical links within and between the tests?
- Did the plan include key safety information throughout the process?
- Were icons used effectively and consistently?
- Were the display diagrams clear and accurate?
- Could an outside observer easily interpret all diagrams?

Reviewing the Process

At the close of your Petroleum Refinery Technician training, reflect on the experiences you have had. Discuss the following questions with other members of your class:

- What have you learned about the importance of petroleum refining in the modern industrial world?
- What roles do refinery technicians play in making high quality and affordable petroleum products available to the world?
- What have you learned about the properties of the components in a crude oil mixture? What are some of the challenges in separating those components?
- Why do you think so many tests must be performed before crude oil is ready for refining? What would happen if these tests were not performed?

- Which of the tests were the most difficult to perform, and why? Which were the easiest, and why?
- What elements of your technician training were most effective for learning the basic test procedures?
- What suggestions would you make for improving the training?
- What opportunities for training are available in your area? What are the job prospects for trained technicians?
- Would you consider becoming a petroleum refinery technician? Why or why not?

glossary

Asphalt: By-product of the leftover hydrocarbons from the cracking process; the material is mixed with rock and gravel and used for paving.

Boiling point: The temperature and pressure at which a liquid changes to a gas.

Carbon: An element that is present in all organic materials; it has four outer electrons that form covalent bonds to other materials.

Catalyst: A substance that speeds up a chemical reaction rate without being involved in the reaction.

Centrifugation: A process used to separate materials by differing densities.

Corrosion: A reaction that causes the deterioration of a substance.

Covalent bonding: A type of bond between atoms; the outer electrons are shared.

Cracking: A process of converting crude oil by breaking apart the large molecules into smaller more usable molecules.

Crude oil: Petroleum that is pumped directly from the ground.

Desulfurization: A process used to remove the sulfur from petroleum; it is often called "sweetening."

Distillation: A process of separating compounds based upon their boiling points.

Double bonds: A covalent bond between two atoms; four electrons are shared.

Engine knock: The premature combustion of gasoline in an engine; this combustion wastes energy and may damage an engine.

Fractions: A mixture of hydrocarbons with similar boiling points.

Gas chromatograph: A device used to analyze organic compounds by heating a material into a gas phase and separating it from other materials.

Hydrocarbons: A group of compounds (including alkanes, alkenes, and alkynes) composed of only carbon and hydrogen atoms.

Octane: The measure of the anti-knock properties of gasoline.

Oil refinery: A facility where crude oil is converted and separated into fractions based upon boiling points.

Organic molecules: Compounds that contain the element carbon (although not all carbon-containing molecules are organic): carbon dioxide is an example of an inorganic molecule that contains carbon.

Petrochemicals: Organic chemicals made from petroleum.

Petroleum: A mixture of organic compounds mostly composed of hydrocarbons.

Pipeline: A series of pipes used for the transportation of petroleum and petroleum products.

Pour point: The lowest temperature at which a substance will flow.

Specific gravity: The ratio of a material's density to the density of water.

Vaporization: The process of changing a liquid into a gas.

Volatility: The ability of a substance to become a gas based upon vapor pressure and temperature.

additional resources

Websites

1. *http://siri.uvm.edu/msds/;* Vermont SIRI MSDS Archive—Site One. Searchable database of MSDSs for many products.
2. *www.xmission.com/~nef/;* National Energy Foundation, 5160 Wiley Post Way, Suite 200, Salt Lake City, UT 84116; 801/539-1406. Variety of educational resources, including colorful, inexpensive posters on petroleum, lesson plans, glossaries, and much more.
3. *www.api.org;* American Petroleum Institute; 1220 L St., NW, Washington, DC 20005. Variety of educational materials, including a video, "Fuel-Less"; briefing papers; and links to other petroleum-related sites.
4. *www.chevron.com/explore/science/crude/;* Chevron Learning Center. Information on crude oil.
5. *www.osha-slc.gov;* Occupational Safety and Health Administration; U.S. Dept. of Labor. Safety guidelines and science background on petroleum refining.

credits

Cover photos:	chip: David C./CORBIS; pills: Jean-Pierre Lescourret/CORBIS; vials: Tek Image/Science Photo Library/Photo Researchers, Inc.; police tape: PhotoDisc, Inc.; cows: PhotoDisc, Inc.; pipeline: SuperStock, Inc.
p. i (right):	Craig Aurness/CORBIS.
p. i (bottom and repeated throughout):	David Zaita/Photonica.
p. v:	Michael St. Maur Sheil/CORBIS.
p. ix (top and throughout):	Dynamic Graphics, Inc.
p. xx (top):	Corbis, Inc.
All other photos:	American Chemical Society/Take One Video.